636.083

£4.95

DMU 0367596 01 5

636083
MCN

KT-487-920

LIVESTOCK HUSBANDRY TECHNIQUES

J.I.McNITT

Lincolnshire College of Agriculture and Horticulture, Riseholme

INSTRUCTIONS TO BORROWERS

Please replace on the correct shelf any books you take down for use in the library room.

Books taken out

Enter your name and date on the card, and place the card in the box provided.

You may keep the book for a fortnight, after which it should be returned.

Return of borrowed books

These should be returned to the Enquiry Office or placed in the slot in the library and NOT replaced on the shelves.

Granada Publishing Limited – Technical Books Division
Frogmore, St Albans, Herts AL2 2NF
and
36 Golden Square, London W1R 4AH
117 York Street, Sydney, NSW 2000, Australia
100 Skyway Avenue, Rexdale, Ontario, Canada M9W 3A6
61 Beach Road, Auckland, New Zealand

Copyright © 1983 by J. I. McNitt

British Library Cataloguing in Publication Data
McNitt, J. I.
Livestock husbandry techniques
1. Livestock
I. Title
636   SF61

ISBN 0-246-11871-7

First published in Great Britain in 1983
by Granada Publishing Limited

Printed in Great Britain by
Richard Clay (The Chaucer Press) Ltd
Bungay, Suffolk

All rights reserved. No part of this publication may
be reproduced, stored in a retrieval system, or transmitted
in any form, or by any means, electronic, mechanical, photocopying,
recording or otherwise, without the prior permission of the
publishers.

Granada ®
Granada Publishing ®

Contents

Preface

Textbooks of animal husbandry which describe such techniques as castration, docking, drenching or dehorning generally assume that the operator knows how to drive the herd or flock to the place of work and how to confine or cast the animal properly once it is there. Such an assumption may have been valid in the past in many parts of the world where such information and techniques were already known by the students since the majority came from farms and had previous experience. With increasing urbanisation, however, more and more of the students undertaking courses in livestock production have not had such previous experience. It is also true that such experience may have been based on inefficient practice, and wrong for the welfare of the animals, and the safety of the operator.

This book has been written to meet the need for a text which describes such basic techniques as how to drive a herd or flock, how to use and read a rectal thermometer and to load a pig onto a lorry. The aim has been to provide a working text for a course in livestock techniques to be taught in the first year of the college or university curriculum in which the techniques will be demonstrated and the students given practice in a practical situation. The material should also be of use to the stockman with some experience as the sections are intended to be self-explanatory; so anyone with some prior experience will be able to carry out the procedures on his own.

Although the primary aim in writing has been to provide material which is not readily available elsewhere, no attempt has been made to provide an exhaustive description of all the methods that can be used. The methods included are generally those most commonly adopted or those favoured by the author. Since the book is intended for use in both developed and underdeveloped areas, consideration has been given to meeting equipment requirements inexpensively since farmers in many parts of the world do not have sufficient capital to purchase expensive tools which are only occasionally used. Thus, the use of water pipe or electrical conduit has been included as an alternative to a disbudding iron and the use of an automotive brake cable for dehorning as an alternative to embryotomy wire.

The author wishes to express his appreciation to all who have contributed to the writing of this book. This includes the several typists who have typed the several drafts, the persons who have contributed drawings and the many people who have made various suggestions regarding content or who have pointed out alternative techniques or pieces of equipment.

J.I. McNitt.

Introduction

Livestock are animals which are kept on a farm for productive purposes. These may mean the sale of meat, milk, eggs or wool, or may concern the use of these stock products at home. In addition to these saleable commodities, other products such as hides, manure, draught and social prestige can be included when livestock are kept for home use.

Since livestock are kept for their products, good husbandry demands that they are cared for in such a manner that they will produce to the highest possible level with the management available. This means that the housing and feeds should be those which will provide for optimum production. As long as the needs of the animals are met, these need not be lavish nor extensive as excess will increase the costs of production and may, in fact, be counter-productive. For example, feeding a dairy cow so much that it becomes fat does not make economic sense and is neither good for the cow nor its productivity.

Production will also be affected by the interactions of the animals with their stockmen. It has been shown that dairy cows give more milk, pigs grow faster and chickens produce more eggs and meat when they are managed by good stockmen who take the time to work with and observe the animals and who really like the animals with which they work.

CHARACTERISTICS WHICH AFFECT MANAGEMENT

Animals are living beings which are not only aware of their surroundings, but also react to them. If the surroundings are upsetting or aversive the reactions may be negative. If the surroundings are favourable the reactions will generally be positive. The word 'surroundings' is used here to encompass all aspects of the space which the animal occupies; housing, other animals, feeds, water, people, management, etc.

Stress is a very important aspect of the physiological functioning of animals. The general effects of stress can be likened to the changes which occur when a person becomes angry. The blood pressure rises, circulation and respiratory rates increase and nutrients are shifted to the muscles for physical action. The reaction may be much stronger than this or may be much weaker. In fact, stress is continually placed on all animals either from within (such as hunger or sex drive) or from without (reaction to other animals in a fighting or mating situation, changes in the weather, etc.). As such, we must consider stress to be a normal part of the lives of our domestic stock. Some stresses, however, come from outside as a direct result of management. We shall refer to these as distresses. These directly interfere with the welfare and productivity of the animals and should be avoided or overcome as much as possible.

One method of avoiding distress is early training. This may involve training dairy animals to enter the milking area, training oxen to be led by a halter or nose ring, or teaching calves or lambs to move easily through a crush. The most important aspect of early training is getting the animals used to people. Domestic livestock will be handled by people all their lives and if they can be taught early that such handling is seldom aversive or uncomfortable and often is pleasant (such as being fed), the animals will be much easier to work with.

Training animals to be unafraid of people is perhaps the most important aspect of stockmanship. A good stockman works not only with the younger livestock but with all. This involves regularly moving through them, talking, stirring the feed and, perhaps, occasionally patting one. Such contact with the animals allows the careful observation for any which are abnormal, the talking calms them and stirring the feed gets them up and eating - all to the stockman's advantage.

Of all the above techniques, talking to the animals is probably the most difficult to learn. At first it seems strange to talk to animals when you know that they will not respond. Friends in the vicinity may also make disparaging comments about your sanity but the benefits far outweigh any such negative aspects. It does not matter what is said or in what language as long as a quiet soothing tone is used. This has a calming effect which will be extremely useful at times when the animals are upset by external influences or by management procedures.

In addition to his desire to make his stock easier to handle through his management techniques, the husbandman must always bear in mind that he has a responsibility to the animals. These animals are confined, so it is his responsibility to provide all their needs including suitable housing, clean, fresh feed in adequate amounts, clean, fresh water in clean containers and freedom from disturbing outside influences and predation. Today's domestic livestock are very different from their wild ancestors and, because of selective breeding by farmers, are often unable to exist without human care. Since man has created this dependence, it is his responsibility to care properly for his stock. In some instances, such care may in itself be distressful. For instance, calves are castrated to reduce injuries from fighting, chickens are dubbed to stop other birds from pecking at their combs and pigs are docked to prevent other pigs from biting them. These operations are distressful but less than the consequences if they are not carried out.

GENERAL PRINCIPLES FOR CARRYING OUT DISTRESSFUL MANAGEMENT PROCEDURES

As has been previously mentioned, it is sometimes necessary even with good management to inflict distress upon domestic animals whether it be for the good of the individual animal, for the other animals or for the enterprise. When such distressful procedures are to be carried out there are three principles that must be borne in mind to help reduce the effects on the animals and subsequent production:

> When carrying out such procedures the stockman should aim to minimise the stress upon the animals and the workers involved.

> These procedures should be carried out in such a manner that the animals will not be injured.

> These procedures should be carried out in such a way that the workers are not injured.

In order to meet these objectives, the stockman should ensure before he starts that he has the proper equipment necessary for the procedure and that the equipment is clean and in good working order. The stockman should know how to use the equipment properly, efficiently and quietly. If he has assistants, they should also know the procedure, know what part is their responsibility and be sure that they can carry that out without fear. Fear is perhaps the most dangerous emotion when working with livestock since a frightened person seldom considers the consequences of his actions. The fear may also communicate itself to animals and get them upset as well, thus making handling more difficult.

It is very often necessary to restrain an animal before the desired procedure can be carried out. (It is very difficult to milk a cow that is running about in a field!) Methods of restraint for the individual species are treated in the relevant sections of this book but there are three principles that apply regardless of the species:

1. The animal must feel restrained

Restraint does not necessarily mean that the animal is totally unable to move. It must, however, be confined to the extent necessary for the procedure. Milking a cow may require only tying in a stall. To castrate a bull, it will be necessary to use a bale which will closely confine the animal or it may be necessary to cast it. It is seldom necessary or possible to confine an animal to the extent that it is totally unable to move. The animal, however, must think it is unable to move or it must be restrained in such a manner that it will not injure itself or the workers if it does move. Over-restraint may, in fact, cause problems because the animal may struggle more than with lesser restraint. This may also provide the animal with leverage which could result in breaking the restraint. For example, if a sheep is sitting on its rump with its legs in the air, kicking will make little difference since there is nothing to kick against to give leverage. If, however, the legs are held, the sheep may kick free as the restraint on the legs gives something to push against.

2. The animal must be comfortable

If it is in an awkward or uncomfortable position it may struggle to get more comfortable. This obviously will make the operation more difficult so it is easier to make sure the position is right before starting.

3. The handler must be comfortable

This obviously only applies when a person is physically holding a chicken, sheep, calf, pig or cow's head. While it is intended that the various procedures with livestock be carried out as quickly as possible, there are times when things do not go as planned and a person may end up holding the animal for some time. If that occurs, holding is safer and easier if the holder is comfortable from the start.

SUMMARY

Stockmanship is an art, but it is an art which can be learned. Learning requires patience, observation and thought about what is seen and how it relates to the animals. Basic principles of good stockmanship are the same for all animals but the needs and behaviour vary among species so one person is seldom an expert with all. There are too many differences, some of which may be very subtle, to allow this. It is this need for a long period of close contact requiring careful observation which makes it absolutely necessary that one works with a species that can be truly enjoyed.

The rest of this book provides information on animals, their requirements and handling. Bear in mind that the animals with which you are working are producing for you using the feed and facilities that you provide. It is therefore necessary to provide the best, most skilful care possible and to remember that the animals are aware of and able to respond to their surroundings. You, the stockman or husbandman, have primary responsibility for their welfare.

1 Livestock Facilities

1.1 WATERTANK CONSTRUCTION AND USE

Provision of adequate supplies of clean drinking water is essential to the success of any commercial livestock operation. Thirsty stock do not eat as much food as those which are well-watered; reduced feed intake may lead to a reduction in the animal's production. This water requirement is particularly important in the case of dairy cattle since milk is over 85% water.

Under traditional farming methods, stock are normally watered once daily from rivers, dams or pans. This is poor management because, in the first place, once daily is not often enough. Furthermore, if large numbers of stock are driven to the same point along a river each day, the continual tramping by the cattle can lead to severe soil erosion problems. Watering stock from the still water of pans or dams greatly increases the hazard of liver fluke infestation (see also Section 5.5) and may result in silting of the dams.

It is preferable that, wherever possible, water be provided from specially constructed tanks which can be kept clean and to which the water supply can be regulated. Every fenced pasture should have a supply of water for the stock. Otherwise, it is useless or, at the least, very difficult and expensive to utilise properly.

When constructing or installing water tanks careful

consideration must be given to their siting. For reasons
of economy, a tank which supplies 2, 3 or even 4 pastures
is preferred to a separate unit for each pasture. The
tanks should be located such that overflow water will run
away from the tank. This is an important consideration
because no matter how well designed and supervised the
tank is, there will be times when the tank overflows.
Livestock also tend to cause spillage when they drink. If
drainage away from the tank is poor, a mud hole will result leading to a health hazard, particularly foot rot (see
Section 2.17). In many cases it may be justifiable to install a concrete apron extending at least two metres out
from the tank. This does not overcome the necessity for
drainage, however; it merely moves the problem further
from the tank.

Since the stock will spend a large part of the time
near the water point, provision of shade - either from
natural vegetation or from specially constructed shelters -
is advised.

Tanks may be constructed from concrete or steel. Concrete tanks should be built on 150 mm footings with 150 mm
thick reinforced walls. Alternatively three sheets of
steel roofing may be riveted together and bent into a 3 m
circle which is then set 75 mm into a 100 mm plinth. The
joints between the sheets should be soldered to make them
watertight. In addition, one may buy, or have fabricated,
steel troughs which have a steel bottom and are thus portable and can be moved as the water requirements change.

However the tank is constructed, a drain in the bottom
of the tank should be provided to facilitate cleaning.
Also a chamber to protect the valve is strongly advised to
prevent damage by the stock or from vandalism. This
should be constructed in such a way that a padlock can be
used to prevent tampering. Be sure that the padlock is
not submerged in the water when it is locked in position.
Metal tanks must either be galvanised or painted to prevent rusting. Lead-based paints should be avoided as the

lead is poisonous to livestock.

The height of the water tank depends greatly on the livestock to be watered. If the sides are too high, the stock may have trouble reaching the water in the tank but if the sides are too low the animals may walk in the tank or foul it with faeces or urine. Table 1 gives recommendations for trough heights and space requirements for livestock.

A mature cow requires 32-36 litres of water each day. Since cattle tend to move and water as a herd, the demand upon the water supply will be very great at the times the cattle are drinking. The problem of depleting the tank can be overcome if the watertank is supplied by a large bore 25-50 mm pipe from a supply tank as shown in Fig. 1.

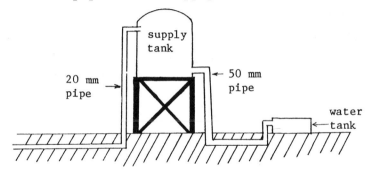

Figure 1. Schematic diagram of a water supply to prevent depletion of water available to stock.

The supply tank should be large enough to hold a three day supply of water for the entire herd. Both the water and the supply tanks should be controlled by automatic valves to prevent low water levels or overfilling.

Table 1 shows average water consumption for the various species. It should be borne in mind that these are not set figures and will vary according to the size of the animal, the nature of the feed being eaten, the productive function of the animal, the ambient temperature, the temperature of the water, the frequency of watering, the availability of the water, the quality of the water and, last but certainly not least, individual variation.

Table 1 Water: consumption and space requirements.

Species	Average daily consumption (litres/day)	Trough space required per animal (mm)	Trough height (mm)
BOVINE			
Mature	32 - 36		610
Dairy cows	(4 - 5 1/1 milk)		610
Yearlings	25 - 40		500
Calves	15 - 25		460
OVINE			
Mature			460
Dry ewes	8		
Ewes with lambs	11		
Rams	11		
Lambs	2 - 4		
PORCINE			
Sows - dry	4.5 - 9.0	600	305 - 380
- lactating	18 - 23	600	305 - 380
Boars	9	600	305 - 380
Baconers	4.5 - 9.0	300	300
Pigs (4-5 weeks)	4.5	225	230
AVIAN			
Layers	0.2 - 0.4 (2 - 3 1/kg dry food)	50	At the height of the tail of the average bird in the pen
Broilers	0.10 - 0.15	25	
Pullets	0.15 - 0.20	25	
Chicks			
- up to 2 weeks	0.08 - 0.11	10	
- after 2 weeks	0.08 - 0.11	15	

If continuous water supplies cannot be provided, stock should be given or taken to water at least twice each day. Dairy cattle in milk should be watered more often.

Proper water supply management is an important part of

stockmanship. All water troughs should be checked regularly to see that an adequate supply of clean water is available. The tanks should be cleaned regularly to prevent the growth of algae. A good rule to follow to know whether a tank needs cleaning or not is whether you would want to drink the water. If you don't want to drink it because it is too dirty, that is very likely true of your livestock as well.

1.2 FACILITIES FOR FEEDING CONCENTRATES, SILAGE AND HAY

In most parts of the world, good livestock husbandry demands that at some time of the year supplemental hay, silage or concentrate be fed. This feeding may supplement pastures during the non-growing season, perhaps to increase the growth rate of young animals (see Section 1.3), to fatten cattle for slaughter, or to increase milk production. Whatever the reason for the supplemental feeding, there are several basic objectives. The animals should be provided with high quality feed in such a manner that all animals have an equal chance at the feed and waste is minimised.

The primary means of ensuring that all animals have a chance at the feed is to be sure there is enough space along the trough or rack. If the animals are fed once or twice daily, there must be enough room to enable all the stock to feed at the same time. If there is insufficient space at the feeder for all the animals those that are more dominant may not allow the subordinate animals sufficient time to feed. If the stock are self-fed, the length can be reduced somewhat but not more than one third of that needed for all stock. The reason for the need for the extra length is that, because of the herding or flocking instinct of domestic livestock, most of the stock will eat at the same time even when self-fed. Table 2 shows the space requirements for feeding concentrates to cattle, sheep, pigs, and poultry. Unless otherwise indicated, the figures are for group feeding i.e. all animals fed at one time.

Losses are minimised by properly designing the feeding unit. For concentrates, the trough should be at such a height that when the animal is eating, its neck is level. The sides of the trough should be high enough to force the animal to lift its head to remove its mouth from the trough. This means that most of the excess of each mouthful will fall back into the trough rather than on the ground. For feeding hay, the rack should be positioned so that the animal must pull downwards to get a mouthful. The rack should also be designed so that the animals cannot pull out more than one mouthful at a time. A ledge under the rack will reduce loss from the excess falling on the ground. Any hay caught by the ledge may be returned to the rack or may be eaten from the ledge.

An alternative design for a hay or silage feeder for cattle or sheep is the tombstone feeder as shown in Fig. 2. The upright concrete slabs are spaced so the neck will go between them but the head must be lifted to remove it. Making the animal lift its head before backing out will reduce wastage due to animals grabbing a mouthful and moving away before it is swallowed.

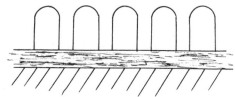

Figure 2. Tombstone feeder for cattle or sheep.

In designing any feeding unit, several factors must be borne in mind.

Build it strong. Livestock do a great deal of pushing and moving about when feeding (especially if there is insufficient trough space) and a poorly built feeder will soon be demolished. Units constructed from timber should be bolted together, not nailed.

Carefully consider the siting of the unit. It should

be in an area which is well drained so that a mudhole does not develop around it.

Provide a concrete apron or build a movable unit.
Even in a well-drained location, the continual tramping of the stock will soon remove any grass cover. Once this happens, any rain will result in a mudhole. This development will cause a reduction in the feeding of the animals and be predisposing to foot problems. This can be overcome by providing a paved apron around the feeder or, alternatively, making the feeder portable so that it can be moved to new ground when its surroundings begin to get muddy.

Some provision should be made for protecting the feed from the weather. Wet feed is less palatable to the stock and may cause sickness if moulds develop. Even if the feed does not become wet, uneaten food should be regularly removed from the troughs.

Space and height requirements for feeding concentrates and hay and silage are shown in Tables 2 and 3, respectively.

1.3 CREEPS AND CREEP FEEDING

Young livestock, especially those still nursing their dams, are highly efficient feed converters. Provision of concentrates to these young animals is thus an economical method of providing the young with a good start. It also will help those whose dams are poor milkers. The rations which are suitable for very young animals are generally highly palatable and more expensive than the rations for more mature stock. Provision must therefore be made for the concentrate to be fed to the young in such a manner that it is not available to the mature animals. This is normally done by means of an enclosure known as a creep which has openings which are too small for the larger animals to gain entrance but allows free access to the young. The size of the opening will obviously vary with the size of the young but should be high and narrow: for calves, 45 cm wide x 90 cm high; for piglets, 20 cm wide x 45 cm high.

The actual design and construction of a creep depends greatly on individual circumstances. It may be an

Table 2 Concentrates: space requirements.

Species	Feeding space required per animal (mm)	Trough height (mm)
BOVINE		
Mature	660 - 760*	560 - 610
Yearlings	560 - 660*	510
Calves	460 - 560	405 - 460
OVINE		
Mature	305 - 380 (group-fed)	250 - 380
	75 (self-fed)	250 - 380
Lambs	225 - 300	200 - 300
PORCINE		
Sows	610 (group-fed)	305 - 380
Boars	305 (self-fed)	305 - 380
Baconers		
50 kg to market	380	230 - 300
25 - 50 kg	305	230 - 300
Up to 35 kg	225	230 - 300
Creep	60	Not over 100
AVIAN		
Layers	75	At or slightly above the height of the tail of the average bird in the pen.
Broilers	50 - 75	
Pullets	50 - 75	
Chicks		
First 2 weeks	25	
Weeks 3 - 6	50	
7 weeks and older	75	

* Add 300 mm per head for horned animals.

Table 3 Hay and silage: space requirements.

Species	Feeding space required per animal* (mm)	Rack or trough height (mm)
BOVINE		
Mature	610 - 760**	760
Yearlings	510**	610
Calves	Up to 450	450 - 380
OVINE		
Mature	460 - 610	305 - 380
Lambs	305 - 380	225 - 305
PORCINE		
Sows and boars (on lucerne hay)	75	

* Feeding from one side only.
**Add 300 mm per head for horned animals.

enclosed area of an existing structure, a small shed or a specially designed feeder which can be moved on skids or on the 3 point linkage of a tractor. No matter how the creep is constructed, it should provide protection from the weather for the feed and, to some extent, for the animal, and must be strongly built as the mature animals will try to enter if they know feed is present.

There are a number of factors which will greatly affect the feed consumption of the young stock. If the pasture is of high quality and is palatable to the stock, less creep feed will be consumed. Consumption will also be reduced if the creep is not properly located. The creep should be located at the place in the pasture where the cattle rest. If the pasture has been well planned, this will also be the place at which shade and water are provided (see Section 1.1). As the young get older, their capacity for food will increase and the milk production of the dam will decrease. As a result, the consumption of the creep feed will increase.

A fourth, and perhaps the most important, factor influencing the consumption of feed is the type of concentrate offered. The ration should be made highly palatable by inclusion of feeds such as molasses or bran, be coarsely ground, and include a wide variety of ingredients. Although the feed should have a high crude protein content (20-24%), unpalatable protein sources such as fish or carcass meal should be limited to 5% by weight.

Some difficulty may be experienced in teaching the young to use the creep. If it is possible, penning the young inside the creep for a day or a night will generally result in their finding the feed and starting to consume it. It is a simple matter to teach the new young if there are older stock around which are still using the creep as the new young will tend to follow the older ones.

1.4   CORRAL DESIGN

Whether one has five or five hundred head of livestock, there will be a need for facilities for confinement and handling for such operations as dipping, sorting, castrating, vaccinations and veterinary treatment. In fact, in order to carry out a large proportion of the techniques described in this manual you need such facilities. These are known by numerous names including corrals, yards, kraals, pens, folds etc. Depending on the complexity of the system it may include some or all of the following: pens of various sorts, crush (chute), dip, sorting gate or shedder, loading ramp, scale and bale.

Pens

There are four sorts of pens normally encountered in a corral system.

The *holding pen* is merely a large pen for confining groups of livestock for variable lengths of time. The pen should have a space of about 1.3 m^2 for each 275 kg animal and 1.6 m^2 for each 550 kg animal. These pens should be provided with shade, water and feed if the stock are to be confined for any length of time. Where stock are confined for long periods in holding pens, they are often construc- ted of cable in hot countries since this allows free move- ment of air resulting in a cooler micro-climate for the stock. Rectangular pens will reduce fighting and upset among cattle that are to be held for some time. This is because the animals prefer to lie along the perimeter and the long, narrow pen provides more perimeter than square pens.

The *forcing pen* is used to funnel stock into a crush. It should never be constructed as a V as shown in Fig. 3(a).

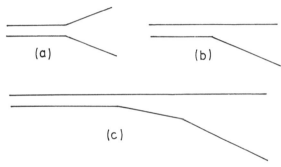

Figure 3. Configurations of forcing pens.

Such a configuration will result in a 'whirlpool' of stock at the entrance to the crush and interfere with smooth stock movement. The preferred construction is shown in Fig. 3(c) although the single constriction diagrammed in Fig. 3(b) is adequate. The key to these is that there is a straight wall along which the cattle can be moved into the crush. When filling a forcing pen, it should never be tightly packed and only a few animals should be started into the crush at one time. Never leave a single animal in the forcing pen as it may get excited and be difficult to start down the crush.

Draining pens (or draining crushes) are normally at the exit side of the dip and allow excess dip solution to drain off the animals and return to the dip supply tank.

Many corral systems will also include *calf pens* into which calves can be sorted for handling and treatment. These are similar to the facilities for mature cattle but are built on a smaller scale for the calves. The fences also have to be tighter as very young calves can get through surprisingly small spaces. Very complex handling systems may also have a special crush for the calves.

Crush

A crush is a long, narrow alleyway through which the stock can move one at a time. The stock can also be confined in the crush by placing poles in front of the lead animal and behind the rear animal. This allows carrying out many practices which do not require close confinement of the livestock. Crushes are often built with some sort of curve or angle as shown below. Since animals are easier to work if they cannot see where they are going, construction of a curved crush with solid sides provides a more even flow of stock. The curve is also an advantage if the crush is to be packed with all the animals' heads on one side since the animals tend to walk with their heads to the outside of the curved crush.

The sides of the crush may be constructed of wood poles, planks or pipe. Wood is preferable to metal as it is quieter and does not frighten the stock as much. The walls of the crush may be vertical or angled. The vertical crush is cheaper and easier to build and there is less tendency for animals to get their legs stuck through the sides. It is also easier to reach animals over the top of the crush with vertical sides than with angled sides. The serious disadvantage is that calves or other small stock may turn around unless extra spacers are provided. The crush with angled sides is self adjusting since the smaller animals are in the lower, narrower portion. The disadvantages of this type of crush are that the animals may try to climb the sides or can easily get jammed and

possibly injured if they lose their footing and go down
in the crush.

The following are suggested dimensions of the vertical
and angled crushes for cattle.

Vertical: 1200 mm high, 610-680 mm wide.

Angled: 1200 mm high, 350 mm wide at ground level
and 910 mm wide at 1200 mm above ground.

A crush for sheep should be 40 cm wide.

Dip

In areas where dipping of livestock is necessary, the
corral system will normally include a plunge dip or spray
race and a foot bath. These are described in Section 1.5.

Sorting gate

It is frequently necessary to separate specific animals
from a herd. This is greatly facilitated by a sorting
gate as shown in Fig. 4. As the livestock move down crush
A, they may be diverted into crush B or C at point D depending on the position of the gate. Instead of crushes,

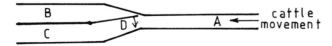

Figure 4. Sorting gate.

B and C may be pens. The gate itself should be at least
3 m long with a solid face. The long gate is necessary so
the cattle can see into the crush they are to enter.
Crush A should be six or eight stock lengths long so the
gate operator has time to decide which way an animal
should go. This is not a serious problem if the animals
are moving slowly and steadily but if they are running,
the man on the gate will need to think and act very quickly if crush A is too short. The operator of the gate
should have a handle so that he can operate the gate from
the side rather than leaning into the crush. In the

latter situation, he will be in the path of movement of the cattle and will impede the flow.

Loading ramp

If livestock, especially cattle, are to be loaded into lorries, it is imperative that a ramp strongly made of wood or of concrete be available. The entrance to the ramp should be preceded by a curved crush so the stock can start moving through the crush and continue in a smooth flow up the ramp and onto the lorry. If the stock are to be weighed before loading, the corral system should be designed so the cattle can be moved onto the weighbridge for weighing and then directly through the crush system to the loading ramp. The side walls of the ramp and the gates connecting to the lorry entrance must be solid so that the animals are not distracted. If the ramp height increases more than 100 mm in each 600 mm of length (over a 10° slope), steps should be installed. Each step should have a 90 mm rise and 300 mm tread width. No slope for livestock movement should exceed 20°.

Scale

The use and operation of the cattle weighbridge is discussed in Section 2.14. The positioning of the scale in the crush is important but depends greatly on the type of scale used. It is desirable that the scale be positioned in a special crush to avoid the cattle having to cross the scale every time they enter the crush for dipping and treatment. Such continual movement across the scale may damage the weighing mechanism and if the livestock have to move through the treatment area to reach the scale weighing will be more difficult as the animals will remember their previous experience and be reluctant to move through the area where this was carried out.

Bale

A bale is a specific area within the crush where one animal can be confined. This may be a complex system with a head gate and restraint facilities or may merely be a section approximately the length of a beast which can be blocked on both ends by poles across the crush. The head gate is designed to catch the animals by the neck. Since both the head and the shoulders are wider than the neck, a properly adjusted headgate will firmly hold the animal. The gate can be formed of two vertical sides which close on either side of the neck or may be a guillotine type with closures which come from the top and the bottom. For straight through work, the former is more common and is often constructed of two poles which are hinged at the bottom and pulled together at the top by means of a rope.

The bale should be built so that the next animal to enter the bale will see the previous animal leaving quietly. When freeing an animal from a headgate, open the gate slowly as the animals tend to fight the head gate if it is rapidly opened. Fighting generally results in movement backward and difficulty getting the animal through. This will also affect the next animal's ease of entry.

When designing livestock handling facilities, you must consider not only the efficiency of moving the livestock and the engineering requirements of the system but also the fact that humans will need to move around through the facility. If such movement involves frequent climbing of fences, the efficiency will be greatly reduced. The necessity for climbing fences can be greatly reduced if, in areas where heavy human traffic is anticipated, gaps 300-380 mm wide are left in the fences. If there is danger of animals getting through such gaps e.g. in calf areas, self-closing, spring-loaded man-gates can be installed. The work area should be provided with a table, a standing rail along the crush to make reaching into the crush easier and some sort of protection from sun and rain.

The last should be of solid construction rather than slats as the slats will, in direct sunlight, result in stripes of shadow falling on the livestock movement areas. This will make moving the stock more difficult as they are reluctant to walk on striped floors or into shadows.

When designing facilities for sheep, the strong flocking instinct should be borne in mind. Because of this, sheep tend to move most easily as groups in wide, straight, flat alleys. The effect of moving from shadow to bright light or vice versa is only on the lead sheep. Once the lead sheep has passed the contrast, the others will follow.

There are several factors to keep in mind when designing a system of corrals.

Be sure your design allows for expansion should this be necessary due to increase in herd size.

Wherever possible, floors should be of concrete with slopes to allow for drainage away from the corrals. The concrete should be tamped, not trowelled, to provide a secure footing as cattle are afraid to move on areas which, in the past, have proven not to give good footing. A brushed concrete surface may leave points which are too sharp. If concreting the floors is not possible, ensure that the system is constructed in an area with good drainage since heavy concentrations of livestock in one area will inevitably lead to mud if there is any rain.

Where machinery is to be used, be sure that gates are wide enough to allow tractors to enter the system for mucking out, etc. A ramp for pushing muck up into a muck spreader may be useful or it may be helpful to provide for muck storage.

Wherever possible gates should be designed to swing in a 360° arc.

Wooden poles should be treated to prevent termite attack and rotting, but care must be taken that the treatments are not toxic to animals. Lead-based paints should be avoided for this reason.

When you have designed your system and have drawn a plan, have several other people check your ideas. It is easier to correct errors on the plans than to reconstruct or modify the system after it is built. Of particular concern should be the movements of people and stock through the system for various purposes.

1.5 DIP DESIGN

Livestock may be dipped using either a spray race or dipping bath. The spray race is more expensive to install and operate but, if properly installed and operated, may give better parasite control than the dipping bath because the dip is freshly mixed each time it is used so there is none of the uncertainty regarding dip strength that there may be with the bath. Also, there may be less danger of injury to the animals when a spray race is used. In some countries such as Great Britain, however, dipping by total immersion in a dipping bath against sheep scab is compulsory.

A spray race is a complex of equipment which includes not only the physical buildings and fences but a pump, motor, nozzles and pipes. Because of the necessity for the proper adjustment and alignment of these components, it is probably best to use a manufactured spray race rather than try to design a race from individual components. Several factors, however, should be considered in the installation and operation of a race.

Build the race so that when the cattle enter they walk away from, not into, the sun. The race should also be built across the direction of the prevailing winds so the winds do not blow through the race. The dip should be sited so that the stock do not have to swim in going to or coming from the dip.

Never build the race near or under trees because falling leaves and seeds may block the sieves and nozzles.

The spray race should be at least 100 m from dairy buildings so the dipping compounds do not contaminate the milk.

The nozzles should be adjusted to deliver 700 litres

per minute at a pressure of 1.4 bar.

Be sure all nozzles are clear before starting to spray.

After use, flush all pipes and nozzles with clear water to prevent buildup of oily residues.

The operation of a dipping bath requires that the animal be thoroughly soaked and wetted to the skin. The length of the bath (i.e. how far the animal has to swim) depends on the type of dipping compound used. Arsenical compounds require longer contact with the animal than the newer organic dips. Tanks for the modern chemicals hold about 15000 litres. There are several features of the design of dipping baths which should be emphasised.

Before the animals enter the bath proper, they should walk through a footbath 3.0 - 4.5 m long and about 200 mm deep to clean the feet and reduce contamination of the dip. The floor of the footbath should contain longitudinal ridges. These will tend to spread the animals hooves as they walk through, thus ensuring thorough cleaning.

The walls of the race from the exit from the footbath to a point about 3 m along the bath toward the exit should be solid and 1.8 - 2.0 m high. This tends to reduce dip losses due to splashing as the animals jump into the bath and also tends to facilitate movement of the animals through the bath as the only exit they can see is straight ahead.

From the footbath, the animals walk up and then start downward into the bath. Some baths are designed with a sheer drop at the point of entrance whereas others have three or four broad steps leading into the bath. These function to force the animals to jump into the bath with the head downwards thus wetting the head and ears. If the animals do not submerge their heads upon jumping into the bath, they should be forcibly submerged as they swim the length of the bath. Whether one uses the sheer drop or the steps is a matter of personal preference but in no case should a ramp lead into the dip. A ramp is ill-advised because many animals will refuse to go further once their feet are in the water. This means that each animal will need to be forcibly pushed into the bath. A roof over the crush at the jump-off point which slopes down toward the bath will encourage the animals to jump downward to the bath rather than outward or upward. This provides better wetting of the head.

The front part of the bath should be curved to avoid a dead space in which the dip compound may collect. It is very difficult to redisperse the compound if such an accumulation occurs. At 2.5 - 3.0 m from the point at which the animal enters the dip a series of steps should be started from the bottom of the dip. These steps should be 0.4 m wide with a rise of 0.15 m. This provides a firm footing for the animal to get out of the dip.

After the animal leaves the dip, it should move along a crush 25-35 m long or enter a draining pen. Both of these should be concreted and sloped so the excess dip drains back into the bath.

Roofing the bath will reduce the dilution of the dip solution by rain, the amount of dirt and dust which is blown into the tank, and evaporation losses. Surface drainage should be away from the tank to avoid inflow of surface water. This can be accomplished by constructing bunds around the dip if the natural slope of the land does not provide proper drainage.

See Section 2.15 for further discussion of the types of ixocides and insecticides available and the management of stock at the dip.

1.6   FENCE CONSTRUCTION

Almost any livestock operation will require fencing of some sort. This may be used for control and movement of livestock, to keep the stock off cultivated lands or as an aid to pasture management. Fences may be constructed from any materials which will suffice to hold the stock. These include live hedges, brush corrals, stone walls, wooden plank, pipe, or the various types of wire. Only the last will be considered here.

Probably the most common wire fence is constructed of barbed wire. Other types of wire include smooth wire, cable (usually used for corrals and other heavy fences) and the assorted types of woven wire including chain link and unit fencing. The least expensive of these is the smooth wire, but it is ineffective at stopping livestock from pushing through. It is often effective to use barbed wire for the lower strands and the smooth wire for the upper strands since barbed wire will not deter an animal jumping a fence (and may injure the animal) but will stop most animals from pushing through. The extra strands of smooth wire above the barbed wire will tend to make the jump more difficult and will reduce the number of animals attempting it.

Chain link (diamond mesh) fence is extremely expensive and is only used in very special instances. The unit fencing is useful for all small livestock as well as calves and adult cattle. Frequently this fence will have decreasingly smaller openings toward the bottom to keep in the smaller animals and to prevent the stock reaching through the fence to graze.

The fence may be supported by wood, steel, stone, or concrete uprights. The wood posts are the least expensive, but must be chemically treated if they are to be used in areas where attacks by fungus or termites are likely. Steel posts are more expensive and tend not to be as strong as wood or concrete posts. They are, however, easy to install using a driver constructed as shown in Fig. 5. The driver is placed with the hollow of the pipe over the post. As the driver is lifted and dropped, it will force the post in the ground.

Concrete posts can be poured quite cheaply and easily if a form is constructed but should be reinforced with wire in each corner when poured. The main disadvantages of concrete uprights are that they are heavy and somewhat

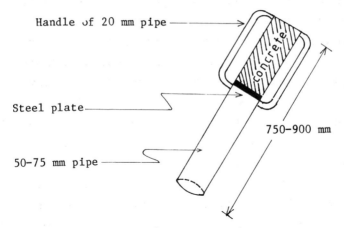

Handle of 20 mm pipe

Steel plate

50-75 mm pipe

750-900 mm

concrete

Figure 5. Driver for steel posts.

bulky to handle. The advantages of concrete posts include resistance to fire, rotting and insect pests.

Posts are normally required every 4.5 - 6.0 m along the fence or they may be placed at 15 m intervals with four supports (droppers) in between. The droppers should not contact the ground. Line posts should be set 460-610 mm into the ground whereas corner posts should be set a metre deep in concrete and securely braced as shown below. Line

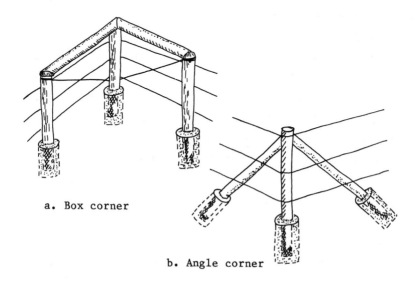

a. Box corner

b. Angle corner

Figure 6. Construction of fence corners.

posts should be in a straight line as the wire will place undue strain on any which are out of line.

The box corner (Fig. 6(a)) provides the most secure bracing but is more expensive and requires more labour than the angle bracing shown in Fig. 6(b). The cross pieces in (a) should be of the same material as the uprights while the tensioners (on the diagonal) should be constructed of No. 10 gauge galvanised wire. The braces in (b) can be the same material as the uprights, or steel posts can be used.

No matter what type of wire fencing you use, it is essential that the strands be tight and securely attached to the uprights. There are a number of commercial wire tighteners on the market which depend either on a ratchet action or involve a block and tackle. When tightening stranded wire, it is important to tighten the top strand first and work downward to the bottom strand because if there is any 'give' in the post, tightening the top strand last may cause slack in the lower strands. When using a commercial wire tightener, each strand should be stressed individually but the same stress should be placed on each strand. An easier method is to clamp a 50 x 75 mm board the full height of the fence on each side of the fence. The tightener can then be attached to the middle of the clamp and all strands tightened simultaneously. The tension on the wire is important as too little tension will result in a slack fence whereas overtightening will result in loss of strength of the wire. Manufacturers' instructions regarding tensions should be followed.

When using staples (U-nails) to attach wire to wooden posts, the staple should be at an angle to the grain so both points do not enter the same grain. When this happens, a small split may occur in the wood and the staple will not stay in. The staples also should not be fully driven home otherwise the wire is difficult to tighten later.

The number of strands of wire to be used depends on

the livestock to be held by the fence. Where only quiet cattle are to be kept, three strands of barbed wire placed 355, 610 and 915 mm above the ground may be sufficient. Four-stranded fences should have the wires at 355, 585, 840 and 1140 mm above the ground and four-stranded fences at 355, 560, 760, 965 and 1170 mm above the ground. Where inadequate feed is available, stronger fences will be required.

It will be necessary to install a gap of some sort in the fence for the movement of livestock and vehicles. It will also probably be necessary to provide some sort of barrier across the gap so that livestock can be kept either in or out of the fenced area. This barrier may take the form of a cattle grid (cattle guard) or some sort of gate. The width of the gap depends on the numbers of livestock and size of vehicles to pass through it.

A cattle grid is merely a pit in the ground over which rails are placed. The intervals between the rails are too wide for the cattle to walk on but close enough for vehicles to cross without problem. The grid should be at least 1.5 m long and wide enough to accommodate the widest vehicles which will regularly be crossing it. A gate should be provided at the side to allow passage of the livestock and any vehicles which may be too wide or too heavy to cross the grid. If the grid is not long enough, cattle may jump over it. The grid should be carefully placed so that silt and soil do not wash into the pit. In any case, provision should be made for removing the rails so dirt which has been blown or fallen into the pit can be removed easily.

A gate may take several forms. It may merely be poles attached horizontally across the opening, it may be four or more strands of wire supported on droppers and a post at the free end which connects to the gate post (concertina gate) or it may be a wooden or steel gate hung on hinges from the gate post. Concertina gates are the cheapest and easiest to build but require constant

maintenance. When a concertina gate is opened, it should
never be laid on the ground. This will lead to tangling
of the strands and injuries to cattle passing through the
gate. When opening such a gate, it should be kept
stretched and carried through a 180° arc and leaned
against the fence. The usual manner of closing such a
gate is by means of a loop of wire at the top and bottom
of the last fence post. The bottom of the gate post is
inserted into the bottom loop and the top of the gate post
pulled to the fence post and top loop dropped over it.
This can be difficult if the gate is tight but can be made
considerably easier if a device such as that shown in
Fig. 7 is used.

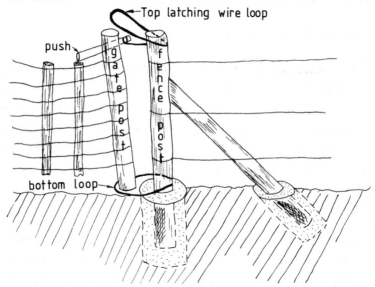

Figure 7. Closer for concertina gates.

To use it, the handle is pulled around the gate post and
pulled toward the fence post. This pulls the gate post to
the fence post so the loop can be dropped over it.

When using a hinged gate, it is imperative that it be
constructed and laid across the opening before the gate
posts are set. In this manner, it will be ensured that
the gap and the gate are the same width. Gate posts require as much bracing as corner posts if not more. The

bracing shown in Fig. 6(a) is satisfactory for swinging tubular steel gates and light wooden gates. If a heavier gate is to be used, more bracing will be necessary.

When hanging the gate, be sure it is placed high enough so it will clear the ground for its full extent. Also, ensure that it will swing shut under its own weight, and that it will swing through 180° so it is out of the way when open.

When building or repairing fences, you will be working with nails, staples and bits of wire. It is extremely important that great care be taken that these are not scattered on the ground as they may lead to foot injuries (see 2.17) or 'hardware disease' (see 2.20).

1.7   ROPES AND ROPE CARE

With the exception of sticks and stones, ropes may have been mankind's first tool. Rope can be made of many things: horsehair, woven leather, animal sinews, papyrus, flax, bamboo, rattan, hemp, manila, sisal and synthetic fibres such as nylon or dacron. Only the last three are currently widely used for rope making.

Rope making

The basic unit of rope is the fibre. Regardless of the material used for making a rope, the single fibre is seldom long enough or strong enough for use in its natural state. The fibres are combined by twisting. A number of fibres are twisted together from left to right to give a yarn. Staggering the fibres within the yarn gives length to the rope while combining a number of fibres together provides strength. Once the yarns have been made, a number of them are twisted together from right to left to produce a strand. Three strands are then twisted together from left to right to produce a rope. This last procedure is known as laying a rope and this type of rope is known as a 'right laid' rope. If the twists are reversed and the fibres twisted from right to left, the yarns from left

31

to right and the strands from right to left, the rope is
known as 'left laid' rope.

Whichever way rope is laid, it is the principle of
opposing twists that makes the rope stable so it does not
unlay easily. Since the twists alternate between the
fibres, yarns and strands, tension on the rope will tend
to tighten the lay since the twists will pull against each
other. The tightness of the lay will also affect the
working properties of the rope. A 'soft laid' (loosely
laid) rope is easy to handle whether wet or dry but has a
low resistance to abrasion. 'Hard laid' (tightly laid)
rope is more resistant to abrasion, but is stiffer and
more difficult to work with. Standard lay is an inter-
mediate rope with medium resistance to abrasion and is
reasonably easy to handle.

Types and properties of rope

Manila rope is made from the fibres of the plant *Musa text-
ilis*. It is a strong, flexible rope.

Sisal rope is made from the fibres of the sisal plant,
Agave sisalana. This rope is only about 80% as strong as
manila rope and is less flexible. Sisal rope is, however,
more resistant to abrasion.

Hemp rope is made from the fibres of the plant *Canna-
bis sativa*. This is a soft fibre which makes ropes as
strong as manila but is generally used only where a soft
laid rope is required.

Of the synthetic fibres, nylon is probably the most
common. It makes a very strong rope with good energy absorption properties and is able to withstand shock-loading.
It is not widely used in livestock work as it is slippery,
thus making tight knots hard to tie and it stretches more
than sisal or manila. For example, if a load of 20% of
the minimum breaking strength is applied to a new nylon
rope, the rope will elongate 25% of its original length.
A manila rope with the same load will only stretch about
10%. Other synthetic fibres used in rope making include

32

polyester, polyethylene and polypropylene.

The breaking strengths for a variety of ropes and sizes are shown in Table 4. Safe working loads for new ropes are normally taken to be one fifth of the breaking strength for natural fibres and one ninth for synthetic fibres. For example, if a sisal rope has a minimum breaking strength of 1,000 kg, the safe working load is 200 kg.

Table 4. Breaking strengths of some common ropes. (Safe working strengths of new ropes are one-fifth or less of the values shown.)

Rope diameter (mm)	(in)	Manila	Sisal	Hemp	Nylon
			Breaking strength (kg)		
6.4	¼	273	218	341	536
9.5	⅜	614	491	727	1,227
12.7	½	1,205	964	1,295	2,250
19.0	¾	2,455	1,964	--	5,136
25.4	1	4,091	3,273	--	9,273

Ropes which are worn or have knots or splices in them have even lower safe working loads. Knots generally reduce the strength of a rope by 50% and splices by up to 20%. The figures in the table are for a steady pull. If loads are to be dropped onto the rope or the rope jerked (as when a cow lunges back against it) the safe working load will be even less.

Since ropes are often sold by weight while requirements are generally specified in length, Table 5 has been included to allow an approximate conversion. Also included in the table are circumferences at the selected diameters since rope sizes, especially large ones, are sometimes specified by circumference.

Care of Ropes

Ropes are equipment which involve an investment and their improper use can increase the cost of operation or may lead to injury if the ropes become unsafe and break.

Table 5. Approximate length of rope of various sizes per kilogram.

Rope diameter		Rope circumference		Metres/kilogram		
(mm)	(in)	(mm)	(in)	Manila	Sisal	Nylon
6.4	¼	19.0	¾	34.5	34.5	37.6
9.5	⅜	28.6	1⅛	16.6	16.6	17.8
12.7	½	38.1	1½	9.2	9.2	10.2
19.0	¾	57.2	2¼	4.1	4.1	4.5
25.4	1	76.2	3	2.5	2.5	2.4

Ropes should be hung in a dry, unheated room which has free air circulation. It is best to hang them on pegs rather than nails as the nails may rust and discolour the rope or can make a sharp kink in the rope. Ropes should not be laid on the floor for storage nor should the coils from the peg be allowed to touch the floor since moisture from the surface may cause rotting of the fibres.

If ropes get wet, they should be dried before coiling for storage. Hang them as uncoiled as possible in a dry room where they will have an opportunity to dry. Dirty ropes can be dangerous as the dirt makes it difficult to see wear. If ropes become dirty, they can be washed with clean water. They should then be allowed to dry thoroughly before storage. Do not use wet ropes, especially those made of natural fibres as they lose 30-40% of their strength when wet. Synthetic fibre ropes are non-absorbent and their breaking strengths are not affected by wetting. Do not use heat to dry ropes.

Ropes are generally made from natural fibres and are thus susceptible to attack by a variety of chemicals. Do not expose your ropes to chemicals if at all possible. If they are so exposed, wash and dry them before storage. Probably the worst chemical to attack ropes in the live-stock situation is the acid in farmyard manure. If your ropes do become soiled with manure or urine, they should be washed before storage.

Do not allow your ropes to kink. Ropes that are

always pulled in the same direction can develop kinks
which may weaken the fibres. If a rope does kink, throw
a twist into the rope by turning it in the direction
opposite to the kink. Kinking can also be avoided by
occasionally turning the rope end-for-end so that the pull
is reversed.

Do not overload your ropes. Use a 5 to 1 safety factor based on the minimum breaking strengths as shown in
Table 4. Although a rope may not break the first time it
is overloaded, the overload may weaken the rope so it will
break at future times even though it is not overloaded.

Use the proper knots. If the knot might have to be
untied under tension, use a slippery knot. It is heart-
breaking to have to cut a new rope or a newly made halter
because someone has tied it incorrectly. This situation
can arise when an animal is tied and struggles until it
falls. If it is choking or liable to hurt itself, instant
release is required.

Regularly inspect your ropes for weak spots. Weak
ropes are unsafe as they may break when you least expect
it thus freeing an animal you thought was securely held.
To check a rope, first carefully inspect the outer surface
for wear and fraying of the fibres in the outer strands
and yarn. You should then inspect the inner surfaces of
the rope by twisting against the lay to open the strands.
A good quality rope should appear bright and clean on the
inside with no dark spots or areas of mildew. No breaking
or fraying of the fibres should be evident either. If a
rope is in a generally bad condition, it should be discarded. If only one area is in a bad condition this may
be cut out and the rope spliced back together. Bear in
mind, however, that this will reduce the safe working load
of the rope.

You must always consider your personal safety when
working with ropes. A loop of the rope should never be
taken around your hand to increase your pull. If the
other end of the rope should happen to be attached to a

cow which suddenly decides to bolt you would be lucky if
you receive only a rope burn on your hand as the rope is
released. If the release were more difficult you could
break bones in your hand, lose some fingers or be dragged
and severely injured. While it is true that there are
times when it is perfectly safe to wrap a rope around the
hand it is better to form the habit of never doing it so
that you are never in danger of forgetting sometime when
it is not safe. If you need to increase your pull on the
rope, pass it around your back and use the friction on
your body as well as your body weight to provide the
needed pull.

When using a long rope, any portion that is currently
out of use should be neatly coiled and hung or placed out
of the way. Do not work with your feet in a tangle of unused rope as this could lead to injuries similar to those
which may occur if the rope is wrapped around your hand.

1.8   KNOTS, SPLICES AND HALTERS

Knots

There are literally hundreds of knots which can be tied in
a rope or used to tie two ropes together. Most knots are
formed using bights, loops and overhand knots in various
combinations. These are shown in Fig. 8. In an underhand
loop, the free end is under the standing part.

When tying a knot, the short end of the rope that you
are working with is known as the 'free end' while the
other, longer end is known as the 'standing part'.

A knot which is tied with a bight in the free end is
known as a 'slippery' or 'slipped' knot. Slippery knots
are used for ease in untying. Probably the most common of
these is the double slippery reef (or square) knot that is
used to tie ones' shoe laces. A slippery knot is not the
same as a slip knot. The latter is one in which the free
end is tied around the standing part and slips to tighten
around a post or other object when pull is exerted on the

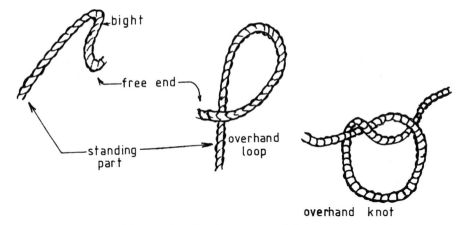

Figure 8. Basic parts of knots.

standing part.

Each knot has a specific purpose and the proper knot should be used for each job. In working with livestock there are four knots which are commonly used: the reef knot, sheetbend, slippery slip (manger) knot and the bowline. While it may sometimes be possible to learn a knot from a written description, it is better to have someone show you the first time. Stockmen and youth club leaders are often skilled with ropes. The descriptions provided here will be most useful in helping you to remember previously learned, but forgotten, skills. Whenever someone is showing you how to tie a knot (or make a splice) it is easier to learn if you stand beside or behind the person to watch than if you stand in front. If you stand in front of the person you are seeing the movements reversed and it is difficult to change them back to the proper procedure in your mind.

Reef knot.   The reef knot (Fig. 9) is used to tie together two ropes of the same size. It is formed of two interlocked bights. To tie a reef knot, form an overhand knot with the free ends, then form another overhand knot with the two remaining free ends. Ensure that the end which was at the back after the first overhand knot

Figure 9. Reef knot.

remains at the back at the beginning of the second. If it
is crossed to the front and the other moved to the rear,
a granny knot with result. This is not as strong as the
reef knot and is more difficult to untie. An alternative
method for tying the reef knot is to form a bight in one
free end and then pass the other free end through the
bight, around both the free end and standing part of the
other rope and back through the bight. If a reef knot is
tied in this way it should be inspected after tying to be
sure that the standing ends of the two ropes are opposite
each other in the knot. If they are diagonally across from
each other the knot is known as a thief's or rogue's knot.
This knot will easily untie if tension is placed on the
two standing ends.

Sheetbend.   The sheetbend (Fig. 10) is also used to tie
ropes together but this knot is used when the ropes are
of different sizes. It is composed of an interlocking
bight and underhand loop with the bight in the larger
rope. To tie the sheetbend, form a bight in the larger of
the ropes. Feed the free end of the smaller rope through
the bight, around both the free and standing ends of the
larger rope and then under itself just where it comes out
from the bight. When the knot is completed, the free ends
of both ropes should be on the same side of the knot.

Slippery slip knot.   The slippery slip knot (Fig. 11) is
used for tying livestock. It is a secure knot which will
hold the animal but it is also easy to untie even with

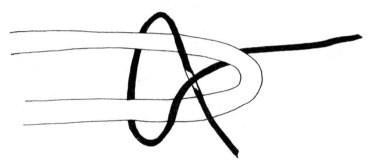

Figure 10. Sheetbend.

tension on the rope. This feature is absolutely necessary
when tying livestock as they may fall or become tangled in
the rope and choke or injure themselves. If this happens
the rope must be loosened immediately; if the wrong knot
has been tied you may have to cut it. To tie the slippery
slip knot, take a turn around a post with the free end of
your rope. Wrap the free end around the standing part and
make an overhand knot using a bight rather than just the
end. Tighten the slip knot around the standing part then
slip it so it is snug on the post. Ensure that the free
end is available for an instant jerk if the knot needs to
be untied. Also be sure that the bight for releasing the
knot is not caught between the free and standing ends as
this may make the knot difficult to untie when it becomes
very tight.

Figure 11. Slippery slip knot.

Bowline.   The bowline (Fig. 12) is formed by interlocking a loop and a bight in the same rope. This differs
from the sheetbend which consists of interlocked loops and
bights from different ropes. The bowline is used to form
a non-slip loop in the end of a rope. This might be for
the first tie when using the Reuff method of casting
cattle (see later) or to form a slip knot for a lariat.

To tie a bowline, form an overhand loop in your rope.
Feed the free end up through the loop. Take it around and
under the standing part (the bottom strand of the loop).
Bring the free end back to the top of the knot and feed it
back through the loop in the direction opposite to that in
which it was first fed. To tighten, hold the two ends of
the bight in one hand and the standing part of the rope in
the other and pull them simultaneously. If you have ended
with a loop in the end of the rope that is too big, feed
the standing part of the bight into the knot and take up
the excess with the free end until the resulting loop is
the correct size.

An alternative method of tying the bowline is to measure along the rope to give the size of loop finally
desired. Form an overhand loop in the rope and pass
through a bight formed in the standing part. Pass the
free end through the bight, around it and bend it back on
itself. Hold the two ends of the bight in the free end in

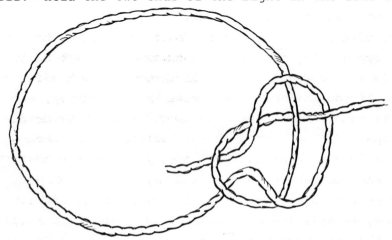

Figure 12. Bowline.

40

one hand and pull the other side of the loop to tighten.

Splices

There are four basic splices which are of importance to the livestockman. These are (i) the end (back or crown) splice which is used to keep the end of a rope from un-laying; (ii) the eye splice which is used to form a perma-nent loop or eye in the end of a rope; (iii) the short splice which is used to rejoin the cut or broken ends of a rope and (iv) the tuck splice used to form an eye in the middle of the rope. Another splice, the long splice, is also used to rejoin a rope but this is not as strong as the short splice and is used only where the rope has to feed through a pulley as the long splice does not increase the rope size as the short splice does.

End splice. The first step in making an end or crown splice is to tie a crown knot (Fig. 13). Unlay about 10 cm of the strands at the end of your rope. As you hold the rope facing you, be sure that the middle strand is at the back. Place a 90° turn in the right-hand strand (as the rope faces you) so it crosses in front of the middle strand. Feed the left strand in front of the free end of the right strand, behind the middle strand, bring it to the front and pass it back through the bight formed when you bent over the right strand.

To tighten the crown knot, hold the rope at the point you stopped unlaying it. Pull downward on each of the strands in succession until all three lay back along the rope. If you have tied the crown knot correctly, the strands should be equidistant around the circumference of the rope. If the knot will not tighten or the strands are not evenly spaced around the rope, you have not tied the crown knot correctly. The crown knot on its own will stop the rope from unlaying. The knot tends to loosen easily, however, so splicing is required. To complete the splice, unlay the rope just below the crown knot and, using any of

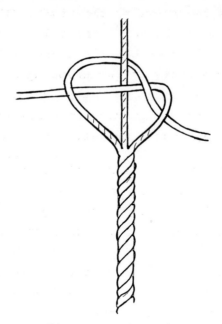

Figure 13.   Crown knot.

the three free ends of the strands, feed it over one
strand and under the next.   Rotate the rope by one-third
and do the same with the next strand then rotate by
another third and tuck under the third strand.   The rule
'over one strand and under the next' is basic to the end,
eye and short splices.   If you are unable to do this and
have to go over two strands or have two strands coming
from under a single strand, you have made a mistake in the
splice.   Continue the 'over one, under the next' procedure
until the ends of the unlaid strands are used up.   Roll
the splice between your hands or on a hard surface to
smooth the splice.

Whipping.   Another method to prevent the end of a rope
from unlaying is to whip the rope.   To do this you will
need about 30 cm of string.   Lay the string in a bight
along the end of the rope.   Hold the bight and wrap the
standing end of the string neatly around the rope starting
from the end opposite the top of the bight.   When you have
used up your string or reached the end of the bight, feed

the free end into the curve of the bight. Then pull on the free end of the bight until the end of the string is under the middle of the wrapping and cut off the two ends of the string. The width of the whipping should be at least equal to the diameter of the rope. Unlike the end splice, whipping does not increase the size of the rope but the whipping may loosen or pull off.

Eye splice.   An eye splice is used to form a permanent eye or loop in the end of a rope. To make an eye splice, unlay about 8 cm at the end of your rope. At the point in the rope at which you want to end the eye, unlay the strands and feed one of the free strands through. Unlay the next strand of the standing part and feed through another of the unlaid strands of the free end. To pass the third unlaid strand, turn the rope around and feed the remaining unused strand of the free end over the unused strand in the standing part and pass it through the unlaid strand toward the front. As with the end splice, if this splice is correctly started the three strands will be facing back along the standing portion of the rope and equidistant around its circumference. To finish the splice feed each of the free strands in succession over one strand and under the next until the free ends of the un- laid strands are used up.

Short splice.   To make a short splice, unlay about 8 cm of strands on each of the ropes to be joined. Marry the ropes as shown in Figs. 14 and 15. The strands of the two ends should alternate. Use a piece of string to tie the ropes together where the unlaid portions join. To com- plete the splice, feed the free ends of the unlaid strands over one strand and under the next until all of the free ends of both ropes are used up.

A good short splice will have 85-95% of the strength of the original rope. It will, however, double the diameter.

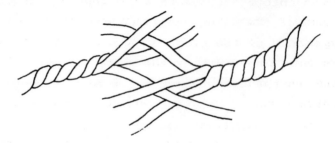

Figure 14. Starting a short splice.

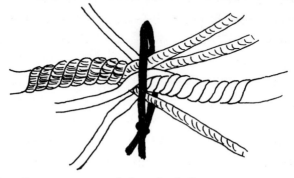

Figure 15. Two ropes married and tied to start an eye splice.

Tuck splice.   This splice is used to form an eye in the
middle of the rope. At the point where you want the eye,
lift one strand of the standing part and pass the free end
through. Adjust the eye to the proper size then lift one
strand in the free end next to the eye and pass the end of
the standing part through. The eye is thus formed by
passing each end of the rope between the strands of the
other end. Since this causes a disruption in the normal
lay of the rope and only one strand of each end is used to
lock the splice, these two factors both weaken the rope to
the point that a tuck splice may have a breaking strength
of as little as 30% of the original rope. This is where
a rope halter generally breaks.

Halters

Halters are made to fit over the face and head of livestock to hold them for leading or for tying. Halters are
made of leather, webbing, chain or rope. The first three

types are generally purchased while the last can be made on the farm. A temporary rope halter can be made by tying a loop approximately the size of the muzzle with a bowline. Slip this onto the animal's muzzle and pass the free end around the head behind the ears. If the free end is then passed through the loop, a halter suitable for temporary use is formed.

To make more permanent halters, the eye and tuck splices can be used. Make an eye only slightly larger than the diameter of the rope using the eye splice. Measure the size of the nosepiece so it fits the animal for which you are making the halter. A useful tool for this purpose is your knee. The muzzle of the mature cow will normally be about the size of your knee. When making calf halters use a shorter distance. When the correct nosepiece size has been decided make a tuck splice at that point again making an eye only slightly larger than the diameter of the rope you are using. To put the halter together, pass the standing end through the eye splice and then through the eye formed with the tuck splice. Whip or end splice the other end and the halter is complete.

For use, the halter should always be placed on the animal so that the free end comes out under the lower jaw on the left side. The spliced nosepiece should be over the nose and the headpiece behind the ears. Placing the headpiece behind the horns is insufficient as the halter will then cross the animal's eyes. Placing the halter on the animal should be practised until putting it on correctly becomes second nature; an incorrectly placed halter may not hold the animal (it will loosen and come off) or may irritate it to the point when it becomes unmanageable.

In selecting ropes for making halters, the minimum breaking strength (see Table 4) and the workability of the ropes should be borne in mind. In practice, 6 mm (¼") sisal or manila rope is generally adequate for small calves and 13 mm (½") rope for older animals. This applies, however, only to rope in good condition. Ropes

which are worn or frayed should not be relied upon to hold
animals.

1.9 CLEANING AND DISINFECTION OF LIVESTOCK EQUIPMENT

As a livestockman there are numerous instruments and tools
you will use regularly to prevent or treat disease among
your animals. Examples include thermometers, syringes,
needles, scissors, calving instruments, drenching guns,
knives, emasculators, tattoo forceps, etc. These instruments involve substantial investment, so should be properly cared for. In addition, improper cleaning and disinfection may lead to the spread of disease rather than
prevention or cure. The procedures which follow are used
to prevent spread of disease-causing organisms from one
animal to another and to prevent introduction of outside
organisms which may cause disease or infection.

The first step after using an instrument is to dismantle it and rinse off all loose debris. Then clean it
thoroughly in soap and warm water. This will not kill the
germs on the instrument but will remove the blood, milk,
dirt, etc. which may protect the germs from the disinfectant. Cleaning should be thorough and a brush used
if necessary. Syringes and needles should be flushed with
clean water before dismantling.

After cleaning, the instruments are ready for disinfection. The method or material used will depend on the
instrument itself, the use to which it has been put, and
what disinfectants are available. The following list is
by no means complete and lists only a few disinfectants
commonly found on farms.

For instruments used without breaking the skin e.g.
thermometers, drenching gun, burdizzo, etc:

Dismantle as far as possible.

Wash thoroughly in warm soapy water (for thermometers,
be sure the water temperature does not exceed the
scale of the thermometer).

Rinse thoroughly.

Dry completely. (NEVER store wet instruments.)

Store in a dust- and dirt-free place.

For instruments which pierce the skin e.g. syringe and
needles, knives, emasculators, etc:

Dismantle instruments as far as possible. (Syringes
and needles should be flushed with clean water immediately after use and before dismantling.)

Wash thoroughly in warm soapy water. Be sure to clean
the inside of the needles. Dried blood is nearly impossible to remove although the cleaning wire which
comes with new needles may sometimes work.

Rinse thoroughly in hot water.

Disinfect using either of the following methods:

(a) Boil 6-10 minutes. (Do not boil knives with
wooden handles as it will cause loosening of the
rivets and shrinkage of the handle from the tang.)

(b) Steam in a sterilizer for 6-10 minutes.

Store in a dust- and dirt-free place. If possible the
instruments should be sealed into paper or plastic
packets after drying.

Unless instruments have been stored in sealed packets,
they should be disinfected before use. Rinsing them in
70% alcohol is usually sufficient if they have been
thoroughly disinfected and carefully stored since the
previous use. Alcohol is not reliable as the only disinfectant for instruments, but is useful for pre-use disinfection or to disinfect needles or instruments between
animals when several are being done at once, e.g. for
blood sampling, dosing with injectable worm remedies, etc.
When vaccinating with live vaccines, alcohol should not be
used as it may impair the usefulness of the vaccine.

Although it is generally not possible to prepare an
injection or incision site as thoroughly in the field as

in a hospital, reasonable care will reduce the chances of infection. The area should be carefully washed with soap (preferably germicidal) and water to remove loose hair, scurf, etc. The area should then be flushed with 70% alcohol, iodine, methylated spirit or other antiseptic. Do not apply the antiseptic with a cloth or sponge. Instead, pour or spray it onto the area. The operator's hands should also be washed and rinsed with antiseptic and the syringe or other instrument assembled without touching the part to go into the animal. For example a syringe and needle preferably should be assembled using forceps but if these are unavailable, only the mount portion of the needle should be touched with the fingers.

As soon as you finish using the instrument, it should be thoroughly rinsed to remove loose debris and left to soak until washing. Washing should occur as soon as possible after use.

Care must also be taken that the crush or pen used to confine the animal is kept clean and any discharges or other material destroyed. Suppose, for example, that you have used a crush to confine a cow while opening an abscess and the discharge had got onto the bars of the crush and the ground. After the animal has been removed, as much of the pus and other tissue as possible should be collected and burned. The crush and ground surface (if washable) should then be scrubbed with hot water and soap and treated with disinfectant to kill any germs present. For this, steam under pressure is best but is not often available. Creosote compounds such as Jeyes' Fluid, or formalin in a 0.5 - 1% solution are both effective disinfectants as long as the area has been well cleaned and is well soaked with the disinfectant. Most disinfectants are more effective when applied in a hot solution. Disinfecting of fences and soil surfaces can be done with dry heat from a blow torch or a welding torch. Sunlight can also be an effective disinfectant as long as the area is thoroughly cleaned and is allowed to dry while exposed to direct rays from the sun.

2 Livestock in General

2.1 USE OF A CLINICAL THERMOMETER

The clinical thermometer is an important tool in maintaining livestock health. While the syringe is primarily for therapeutic purposes, the thermometer is diagnostic. As such it is used to identify and diagnose animals which are sick so that correct treatment can be administered.

Principle of the clinical thermometer

Essentially, the thermometer is a hollow tube filled with a liquid which expands when heated and contracts when cooled. The clinical thermometer (Fig. 16) is filled with a liquid metal, mercury, which has a high coefficient of expansion. It differs from ordinary thermometers in that a constriction at X at the top of the bulb prevents the downward movement of the mercury unless it is shaken down.

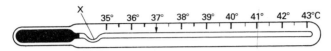

*Figure 16. Clinical thermometer
(from Nelkon: Principles of Physics (1981)).*

Since the mercury will move to the same level each time it is heated to a certain temperature, it is possible to calibrate the thermometer so that an unknown temperature can be determined.

When using a thermometer, always check whether it is calibrated on the Fahrenheit or Celsius scales and know the value of each division before you insert it into the animal. Normally the thermometer will have large lines for each degree and small lines for each 0.1°C or 0.2°F.

In cross-section, the thermometer is shaped like a raindrop, i.e. ⌂ . The scale is marked on one of the flat sides. For reading, you should hold the thermometer horizontally while rotating it slightly with the more pointed side toward you. At one point you will be able to see the mercury column. Check the top of this against the calibrations and you can read the temperature.

Use of the clinical thermometer

The thermometer will almost always be inserted into the rectum. Before insertion it should be shaken down and checked, then moistened with water or a neutral oil. Insertion should be done gently as the lining of the rectum is tender and can easily be ruptured.

For mammals:

> Hold the thermometer with the bulb pointing toward the backbone.

> Insert the thermometer gently bringing it parallel to the backbone after inserting about 10 to 20 mm.

> Insert all but about 20 mm of the thermometer into the rectum.

> Hold in the rectum for 1 to 2 minutes.

> Remove and read.

> Clean and disinfect (see Section 1.9).

For poultry:

> Insert the thermometer 10 to 20 mm into the cloaca.

Hold until the mercury stops rising.

Remove and read.

Clean and disinfect (see Section 1.9).

Normal temperatures.   Table 7 shows the normal rectal temperatures for the common domestic species. It must be strongly emphasised, however, that individual variations in temperature do occur so an animal which does not have exactly (within ± 1.0°C) the rectal temperature shown is not necessarily abnormal. Factors which influence temperature include:

Age: Younger animals often have a higher temperature than adults.

Sex: Females may have a higher temperature than males.

Breed (size): Larger types tend to have lower temperatures than smaller.

Time of day: Temperature generally reaches a maximum in the evening and a minimum during the morning hours.

Feeding: Digestion of food will raise body temperature.

Drinking: Ingestion of cold water may reduce body temperature.

Exercise: Body temperature will usually be increased by exercise.

External temperature: Body temperature tends to follow external temperature to some extent so the body temperature may be higher on hot days than on cool.

Excitement or fear: Either of these may lead to an increase in temperature.

Pain: This also may cause a temperature increase.

Death: Just before death the temperature normally drops to subnormal levels.

Disease: With febrile diseases, the body temperature will be elevated.

Study of the temperatures in Table 6 reveals that, for cattle, one should expect a temperature of about 38.5°C whereas for the smaller animals 39.0 to 39.5°C is about normal. The causes of variation in relation to these facts are what should be remembered rather than attempting to memorise the entire table.

Table 6. 'Normal' temperatures of domestic animals.

Species	°F	°C
Calf	103.1	39.5
Cattle (mature)	101.3	38.5
Chicken	105.4	40.8
Goat	104.0	40.0
Lamb	103.1	39.5
Pig	102.2	39.0
Piglet	103.6	39.8
Rabbit	102.7	39.3
Sheep (mature)	103.1	39.5

2.2   USE OF A SYRINGE

The syringe is very commonly used in livestock operations to administer vaccines to induce immunity to disease or for therapeutic purposes to administer antibiotics, vitamins, hormones, etc.

There are three primary routes of administration using a syringe: (i) subcutaneous (under the skin), (ii) intramuscular (into the muscle) and (iii) intravenous (directly into the blood system). Intravenous injections provide the fastest response to an infection. Subcutaneous injections are the slowest. The intramuscular route is probably most used because the technique is much simpler although the response is slower than by the intravenous route. Subcutaneous injections are used mostly for vaccinations where slow absorption is desirable. For specialised purposes other routes such as intradermal, intraperitoneal and subconjunctival may be used. Ideally all injections should be done by a veterinarian but in practice this is not always possible. The intravenous and more specialised routes certainly should not be attempted by non-veterinary personnel.

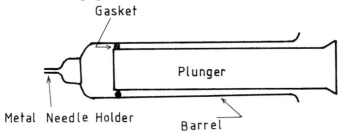

Figure 17. Parts of a syringe.

There are many types of syringe; the one shown in Fig. 17 is the most common. Some syringes have ground glass plungers which require no gasket. Gaskets are necessary for nylon and plastic syringes. The 'push-on' or 'Record' needle mount is shown above. You may also encounter the 'Bayonet' or 'Luer-Lock' mount or a screw-on mount.

Syringes are commonly available in 1, 2, 5, 10, 20 and 50 ml sizes although other sizes are also available. You should always select the syringe which is closest in volume (above) to the injection you plan to give to ensure accuracy of dosages.

Needles have two size designations: their length in millimetres or inches and their diameter (gauge). The larger the diameter of the needle, the smaller its gauge designation. The gauge is normally stamped on the side of the mount of reusable needles. A needle designated as 17G x 1½" is the same length as, but larger in diameter than, a needle designated 20G x 1½" whereas it has the same diameter but is longer than a 17G x ¾" needle. Some needles also carry a metric designation for both diameter and length. Needles from ¾" to 1" are normally used for subcutaneous injections whereas the longer needles (1½" - 2½") are used in intramuscular injections. The length of the needle will depend on the size of muscles of the animal to be injected. Longer needles are normally used for cattle, for example, than for chickens.

How to use

Select the proper syringe and needle. Check the syringe to be sure the plunger moves freely and that the gasket is in place. Check the needle to be sure it is patent (i.e. not plugged) by looking through it or by blowing air through it with a syringe. Attach the needle to the syringe. In the case of push-on mounts, push the needle firmly onto the syringe then give a slight twist to lock the needle on the mount.

Read the instructions on the bottle to ensure you are using the correct preparation in the correct quantity.

Draw air into the syringe in an amount equivalent to the amount of solution you wish to draw from the bottle. For example, for a 4 ml injection you would draw 4 ml of air into the syringe.

Wipe the bottle cap with surgical spirit.

Insert the needle through the rubber stopper in the top of the medicine or vaccine bottle. A needle inserted into a bottle of medicine must always be sterile.

Inject the air from the syringe into the bottle and withdraw the required amount of solution from the bottle into the syringe. You will not be able to fill the syringe unless the tip of the needle is below the level of solution in the bottle.

Withdraw the needle from the bottle. Turn and tip the syringe so the air bubble is under the mount. Gently expel the air bubble from the syringe and needle. If you withdraw too much solution do not return the excess to the bottle as you may contaminate the remainder of the contents.

You are now ready to inject the solution into the animal. Whatever the route of administration, you should be sure the tip of the bevel is against the skin and enters first as shown in Fig. 18. Do not prick the

animal before inserting the needle. Sometimes jabbing
from about 10 mm from the skin surface will improve
penetration.

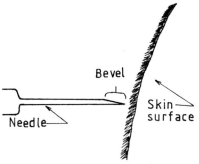

Figure 18. Placement of a needle for injection.

Insert the needle its full depth into the muscle for
intramuscular injections or into the space under the
skin for subcutaneous.

Expel the contents of the syringe smoothly into the
animal.

Withdraw the needle from the animal while holding your
finger over the hole to prevent the injected material
from running out through the hole in the skin. Rub
the area to spread the material away from the injection site.

Disassemble needle and syringe; wash and sterilise
(see 1.9).

2.3 LIVESTOCK IDENTIFICATION - INTRODUCTION

In any progressive livestock operation, identification of
individual animals is absolutely necessary for keeping
livestock records, selection programmes, health programmes,
etc. There are numerous methods of identifying stock, including memory, pictures, eartags, tattoos, brands on the
skin, brands on the horns or hoofs, flesh marks, neck
bands, leg bands, etc. The method of identification depends very much on the needs and the personal preferences
of the stockman and the limitations posed by the livestock.
(Planning to horn brand polled cattle will not aid

identification greatly!) It may also be useful to combine
several methods. For instance, in the case where the
livestock are branded but branding is done only once or
twice per year, it may be useful to eartag the animals at
birth as a means of identification until the next branding session. The cost, the ease of application, the ease
of reading, susceptibility to alteration and permanence
should also be considered when deciding which method to
use.

The most commonly used methods for cattle are pictures
(2.4), eartags (2.5), tattoos (2.6) and brands (2.8 and
2.9). Sheep are normally identified with eartags and pigs
with ear notches (2.7). Poultry are identified by bands
placed around the legs, tags (similar to eartags) attached
to the wings or holes punched in the web of the foot.

Whatever method of identification is used, it is
essential that some sort of numbering system be used. If
you are fortunate enough to be able to establish your own
system, do it with care so that each number will give the
maximum amount of information. Never establish a system
in which a number will be repeated. If, for instance, you
decided to number each calf sequentially as it was born,
starting from the number 1 each year, you could easily end
up with as many as four or five cattle each having the
same number. This situation is not at all conducive to
good record keeping! Be sure also that you leave room in
your system for expansion of the herd or flock.

As an example of a possible system, stockmen sometimes
allow the first digit of the number to indicate the year
of birth and the rest of the digits the sequence number of
birth, e.g. the sixteenth calf born in 1975 would be 516.
It is obvious that this system will work fine until the
101st calf is born. Then it will be necessary to go to
601 which represents 1976 or to use a four digit number.
If the herd or flock will never reach 100 offspring per
year, three digits will be sufficient.

A refinement is to use odd numbers for females and

even numbers for males. Thus the first female born in
1975 would be 501, the second 503, the third 505. This
use of odd numbers continues regardless of the use of the
even numbers. The actual chronological sequence of numbering might thus be 501, 503, 505, 502, 507, 504, 506,
509, 508, etc. With this system it would be necessary to
go to four digits when there were more than about 65 cows
as one must allow for the chance of a very uneven distri-
bution of sexes of the offspring. With ewes allowance
must be made for twinning so the maximum number with three
digits is about 30. With sows it is probably better to
use the litter number in one ear and the individual number
within the litter in the other.

The colour of ear tags can also be used as a part of
the information conveyed representing perhaps the year of
birth or sex. This allows information to be added without
adding extra digits. For listing in record books or
tattooing, the first letter of the colour can be used.
For instance, 506 on a red tag would be R506 or 506R and
301 on a yellow tag would be Y301 or 301Y. Confusion can
arise if similar colours such as red and orange are used
or if the person reading the tags confuses colours.

The basic rules for a numbering system are:

Never repeat the same number on different animals.

Convey as much information as possible with the minimum of figures.

Keep it simple so it will be understood easily by your
stockmen.

Be sure the tags, tattoos, brands, and other equipment
and supplies are readily and consistently available.

2.4 LIVESTOCK IDENTIFICATION - PICTURES

In many cases, the use of pictures of cattle is a useful
means of identification. These may either be photographs
or simple drawings of the markings filled in on an outline

Figure 19. Outlines for drawing on markings.

as shown in Fig. 19.

Drawing markings is not useful for cattle which are all one colour (e.g. Brahman, Jersey, Sussex) and is very difficult if the cattle are speckled and have a large number of small patches of contrasting colour. Line drawings are most useful for breeds like Friesians, Guernseys and Ayrshires which have a basic black or red colour with large white patches overlying the basic colour.

When making line drawings several factors must be borne in mind. Firstly, it will make later use of the pictures easier if you make a positive drawing i.e. shade the areas representing the black or red, not the areas representing the white. Secondly, you will find it is the small or oddly shaped marking which is most useful for identification. For example a black leg in a breed where most animals have white legs will serve as a rapid, positive means of identification. Since these areas are so important, it is necessary to draw them with extreme care. It must also be borne in mind that a young animal is not a miniature of an adult. The relative sizes of the body, limbs and head are different.

Although pictures provide a positive means of identification which is very difficult to alter or obliterate, it is also a very cumbersome method if you must sort through a pile of 100 pictures to identify the one animal you are interested in. It is thus suggested, if you plan to use pictures for identification, that some other method be used in conjunction with the pictures to provide a means of rapid identification.

2.5   LIVESTOCK IDENTIFICATION - EARTAGS

The use of eartags is a quick and easy method of identification which is relatively inexpensive. The disadvantages
of eartags are that they may be lost from the ear and many
of the tags which are available are difficult to read unless the animal is confined in a crush.

An eartag is a piece of numbered metal or plastic
which is fixed by means of a hole through the ear. Metal
tags generally are formed in a U shape with some sort of
locking mechanism for the open end. This may consist of
a small point which is pushed through a hole and bent over
to hold it, or it may be similar to a rivet. Because the
tags form a closed circle they are prone to getting caught
on branches or fences and breaking or tearing out. Plastic tags consist of a large swinging portion on which a
number is written. The tag is fixed by means of spreading
ends on a shaft which is pushed through the ear or there
may be another portion of the tag which is snapped onto
the shaft after it goes through the ear. Though the plastic tags do not form closed circles, they are still lost
occasionally.

The choice of which tag to use depends greatly on the
preferences of the individual stockman but one certainly
should try to obtain a tag which will not easily pull out,
which is simple to insert, which is readable and which has
permanent numbers. Which type of tag to use should be
carefully considered as each type will require a special
applicator and unnecessary expense will be incurred with
frequent changes. It is also advisable to use a commonly
used type to ensure that new tags will always be available.
A source of supply of tags can be, and often is, a real
problem with the less common types. Another consideration
is availability of black tags on which your own numbers
can be stamped (metal tags) or applied with special ink
(plastic tags). The range of numbers normally available

is limited and if you have many stock you will very quickly use all that are available and have to print your own.

Insertion of metal tags

The location of metal tags in the ear is a matter of
choice but it is usually best to have the tag in the upper
portion of the ear with the numbers on the outside. In
this position the numbers on some of the larger tags can
be read from two or three metres away. When inserting
metal tags, follow the manufacturer's instructions for inserting the tag in the applicator. Confine the head of
the animal and close the plier so the catching device for
the tag goes through the tissue of the ear. Be sure not
to pierce the ribs of the ear as this may cause the ear to
droop, and take care to avoid piercing a large blood
vessel. If young animals are being tagged do not put the
tag all the way onto the ear so the margin of the ear
touches the closed end of the tag. As the animal grows
the ear will also increase in size and room should be
allowed for this so the ear margin is not deformed. No
subsequent treatment or disinfection of the wound is normally necessary.

Insertion of plastic tags

Plastic tags are generally much larger than metal tags and
are placed in the lower edge of the ear so the number
hangs below the ear margin. When putting on the Lone Star
type tag which consists of a shaft with spreading ends,
the tag is put into the applicator which consists of a
sharp point with a tube to hold the tag. The point is
then pushed through the ear and withdrawn. The spreading
end of the shaft will cause the tag to remain in the ear.
The other type of tag requires attachment of a locking
flap on the back of the tag. The tag is inserted through
the ear using a hole made with a punch or applicator and
the locking device is pushed into place on the inner end
of the shaft. With plastic tags as with the metal, care

should be taken not to pierce the ribs of the ear. No treatment for the wound is required.

Regardless of the type of tag used it is inevitable that some tags will be lost. It is therefore necessary that another means of identification be used in addition to the tags. It is also important that missing tags be replaced immediately. Waiting weeks or months to replace lost tags may result in total loss of identification since by then there may be several missing tags and deciding which number goes with which animal can become impossible.

Wingbands for chickens are small versions of metal eartags which are inserted in the fold of skin at the rear margin of the wing.

2.6 LIVESTOCK IDENTIFICATION - TATTOOS

A tattoo is a permanent marking which is extremely diffi-cult to alter but requires more effort for installation than an eartag and can often be very difficult to read. A tattoo is applied by making holes in the skin and forcing a marking dye into the wounds. When the wounds heal the dye is retained under the skin and leaves a permanent record of the wounds. For example, tattoo number 903 would appear on the animal as shown in Fig. 20.

Figure 20. 903 as a tattoo.

Although tattoos are normally placed in the ear, it is possible to tattoo almost anywhere on the animal. Other areas often used are the lip and the udder, or the wing web of poultry.

Skin colour may affect the readability of the tattoo, e.g. black ink impregnated into black skin will be very difficult to read. This can be avoided by selecting pink skinned areas or by using variously coloured inks which are available in some parts of the world. Holding an

electric torch behind the ear or holding the ear so the
sun shines through it will also make reading easier.

The inks used for tattooing are specially made so the
defensive mechanisms of the body will not force the ink
out of the wounds. Ordinary writing ink will not work.
If special tattooing ink is unavailable very finely ground
charcoal can be used. This must be an extremely fine powder and must be carefully worked into the wounds made by
the tattoo forceps.

Procedure for tattooing

The following procedure is based on placing a tattoo in
the ear but is similar to the method used on other parts
of the body.

Decide number or symbol to be tattooed.

Place appropriate figures in the tattoo forceps.

Check the tattoo on a piece of paper or cardboard to
be sure the figures are properly aligned. This is
very important and should never be omitted since an
improper mark cannot be changed.

Note the proper positioning of the forceps on the ear
to place the tattoo correctly in the ear i.e. do not
put the number in upside down.

Catch the animal and clean the ear thoroughly with
methylated spirits to remove grease, dirt, etc.

Dry thoroughly.

Liberally apply tattoo ink or paste to the location
on which the tattoo is to be applied. A well-worn
shaving brush is excellent for this purpose. Try to
avoid the ribs and blood vessels of the ear as this
may cause bleeding and/or weakening of the ear.

After properly positioning the forceps, close tightly.

Remove the forceps from the ear and work more ink or
paste into the wounds.

The above procedure is simple and relatively fast, but care must be taken that the complete procedure is followed every time as any minor omissions can result in poor quality tattoos.

2.7   LIVESTOCK IDENTIFICATION - FLESHMARKS

A fleshmark is a cut or hole made in the ear, dewlap or other part of an animal for owner identification or for numbering individual animals. Fleshmarks in cattle are normally used for owner identification and consist of a cut made in the ear, neck, nose or dewlap. The cut may be a hole, a bit of flesh may be left hanging (known as a wattle when on the neck or jaw or a but when on the nose) or the free flesh may be cut off altogether. If each owner has a registered fleshmark and all of his animals carry that mark, it is easy to determine which animals belong to whom. The cuts in cattle are normally made with a sharp knife. In some countries, fleshmarks - other than ear notching - are forbidden by law, so stockmen should check local regulations before including fleshmarks in an identification programme.

Fleshmarks in the form of ear or snout notches are the most common method of numbering pigs. The notches normally are coded so that a notch or a hole in a specific ear or in a specific location on that ear has a set value. Three systems of notching are shown in Fig. 21. This first provides for sequentially numbering the pigs while the second uses one ear to record the litter number and the other the individual pig number within that litter. The third method does not identify the individual pigs but does tell the year and month that they were born. The code used is a matter of preference unless you are involved in a pig production scheme that requires a specific system. If you have a choice of what system to use or want to work one out for yourself you should first decide what information you want to convey. Once you have done this be sure to work out all combinations on paper before

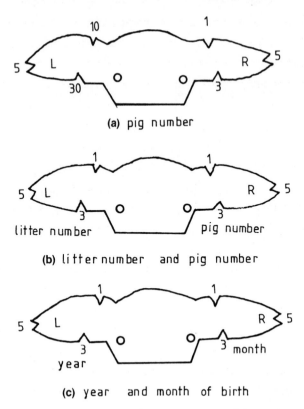

(a) pig number

(b) litter number   and pig number

(c) year   and month of birth

Figure 21.   Three systems of ear notching pigs.

you actually notch any pigs.   Once you are satisfied that
the system conveys the information you want, that a minimum number of notches are used and that all possibilities
within your system are covered, a number of copies of the
system should be made on paper so that everyone working
with the pigs will have access to a code card.   Of course,
people who are reading the notches regularly will soon
learn to read them without reference to a card but personnel who only read them occasionally should be careful
that their readings are correct.

Ear notches in pigs are generally made with a special
tool which cuts a piece out of the margin of the ear or
punches a hole in the middle.   Ear notches can be purchased in a variety of shapes including a V, a crescent,
or a T-shape and punches can be purchased to make a round,

square or star-shaped hole. Notching should be done as soon as possible after birth to minimise bleeding and the check suffered by the piglets. To notch the ears, make sure of the notches you want to make. Place the notcher onto the ear fully to the stop and close the handles. This will cut out the notch. No treatment of the wound is necessary.

It is useful to have a sketch outline of the pig's ears on the permanent record card so the notches can be drawn on. This should be done from the ears and not from memory. In that manner if there has been a notching error it will be recorded and the identity of the animal will not be lost.

Sheep have relatively soft ears and, in brushy country, will often lose eartags. Some flock managers use ear notches for identification to overcome this. The procedure for earnotching sheep is the same as for pigs. If you are switching from eartags to notches, be sure that you have some way to differentiate the notches used for numbering from the tears left in the ears by the tags being pulled out. This will only be a problem for the first few years until all stock that were previously ear-tagged have been culled.

An advantage of earnotching for both pigs and sheep is that there is none of the continuing expense there is with eartags or tattoos. Once the farmer has purchased the notchers there will be no further expenditure until they need to be replaced. Reading notches is somewhat slower than eartags and is more liable to reading error than tags or tattoos.

Fleshmarks in poultry consist of holes punched in the web between the toes. One system for numbering chicks is shown in Fig. 22. The method for punching the toes is similar to that used for earnotching except that a special toe punch is used. Individual identification of chicks is not a commonly practised technique but may be useful to identify particularly good stock.

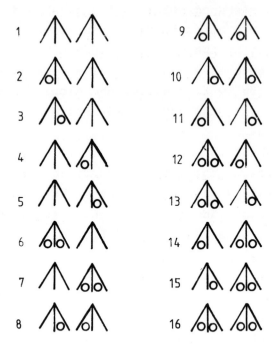

Figure 22. A coding system for toe-punching to identify chicks.

2.8 LIVESTOCK IDENTIFICATION - HOT BRANDING

A brand is an indelible mark that is placed on an animal's
skin. The oldest method for doing this is to use a hot
iron to burn the skin. Use of a very cold iron will also
make an indelible mark as described in the next section.
Sometimes animals are 'branded' with paint, ink or acid.
The first two are not indelible and last only a short time.
Clear brands are difficult to make using acid.

The aim of the hot brand is to cause a wound which
will later form scar tissue upon which hair will not grow.
The skill with which the iron is applied will affect the
quality of the brand. If it is too hot the iron will heat
too wide an area and an iron that is too cool or not held
on the animal long enough will not result in a long-
lasting brand. A disadvantage of hot brands is that the
scar tissue formed after branding reduces the value of the
hide if the brand happens to be in a position from which
valuable leather is obtained. For this reason, brands are

often placed on the hump of zebu cattle or on the lower leg or jaw. The latter positions present problems however as the struggling of the animal may result in a blurred brand as it is very difficult to totally immobilise the leg or head of a calf. Hot brands also cause more pain to the animal than freeze brands.

For hot branding, the animal should be cast and restrained on the ground. The iron, which has been heated in a wood fire to a bluish-white colour is then applied to the hide. The iron should not be too hot or it may start the hair afire, will result in blurring of the brand and will also ruin the iron. If the iron is too hot, lift it and allow it to cool slightly and then return it to the hide. With the iron against the hide, rock it gently up and down and from side to side. This will allow air to get under the iron and give a more uniform brand. It will also ensure that all portions of the mark are evenly burned into the hide. The iron should burn the hair and outer layer of skin. When the iron is lifted the brand should be a dark mahogany colour.

The age of the animal at branding will have no effect on the quality of the brand but young animals are easier to cast and restrain so it is advisable to brand as early as possible. Before branding, the irons should be inspected. If the faces of the iron are thin, it should not be used. A thin iron will cut too deep and will make only a thin scar which will easily be covered with hair and will become hard to read.

Do not brand animals when it is raining or when the animals are damp. The moisture in the hair will cause the heat to spread as steam and will result in a blotched brand, a bad sore or may even give no brand at all. To get the best brands, the hair should be clipped from the brand site. This removes the oil which is normally found in the hair and which may cause spreading of the heat (or which may catch fire) with consequent blurring.

2.9   LIVESTOCK IDENTIFICATION - FREEZE BRANDING

While a hot brand is a scar from a serious burn, a properly performed freeze brand merely causes the hair on the brand site to grow in white. This is because freezing the skin for a definite time will kill the colour-producing cells of the hair follicles (melanocytes) without actually killing the hair-producing cells. If the skin is not frozen for a long enough period, there will be no effect at all whereas if the area is frozen for too long a period, the hair follicles will be destroyed and there will be no regrowth of hair.

The advantages of the freeze brand are that it is permanent and easy to read even when the animals have long hair. This is true because, since the brand is in the hair and not a scar on the skin, the brand cannot be overgrown by the hair. In addition, a freeze brand causes no hide damage so may be placed anywhere on the animal. The primary disadvantage of freeze branding is that it requires very precise timing for the branding with each step in the procedure carefully performed for each brand. It is also more expensive than hot branding.

The 'irons' used for freeze branding should be made of copper. Brass or bronze will work but copper is best. Hot branding irons are unsuitable as they do not have enough metal in them to give thermal mass so the iron will stay cold during the branding process. The size of iron used is a matter of choice. When branding young animals, it is advised that small (25 mm) numbers be used as the brand will increase in size up to six or eight times as the animal grows.

The refrigerant used may be either dry ice (-79°C) or liquid nitrogen (-195°C). With dry ice a solution of 95% ethyl, methyl or isopropyl alcohol or acetone will have to be used so that the irons can be placed in the solution to cool. Note that the alcohol must be 95% because as the water content increases the solution will be less cold and the branding less effective. For branding 20 animals,

5 litres of liquid nitrogen or the same amount of alcohol plus 10 kg of dry ice will be required. Once branding is started all animals should be finished as quickly as possible because the initial cooling of the irons uses a great deal more refrigerant than rechilling between animals.

Freeze branding technique

The hair on the brand site should be clipped as closely as possible to the skin. It is imperative that this be done with clippers and not with scissors as it is almost impossible to clip closely enough with scissors. Removing the hair is necessary to ensure close contact between the iron and the skin. Hair acts as an insulator and interferes with the freezing process. This is especially critical when using dry ice.

Wash and scrub the brand site with alcohol or acetone to remove scurf, dirt and loose hair. After washing, flood the area with acetone or alcohol and scrape off the excess with your hand.

Before starting, the irons should have been cooled in the refrigerant. When the vigorous bubbling of the refrigerant ceases and only a few bubbles are seen rising to the top, the irons will be at the refrigerant temperature. Once the irons are cooled and the brand site is prepared, the iron should be applied to the site for the required time. This time will vary with the age and breed of the animal (thicker skin requires a longer time), the type of refrigerant being used, the pressure of application of the iron, the time of year (during the parts of the year when the hair is growing, the time required will be reduced) and the brand site on the body. Some suggested times are shown in Table 7. The iron should be pressed firmly onto the skin to ensure that all parts of the numeral or letter are in close contact with the skin.

After the iron is removed, a cloth soaked in water at body temperature should be applied to the brand site for

Table 7. Some recommended times for branding cattle.

(a) DRY ICE/ALCOHOL REFRIGERANT

Age	Dairy cattle	Beef cattle
Birth – 1 month	10 seconds	15 seconds
2 – 3 months	15	20
4 – 8 months	20	25
9 – 18 months	25	30
over 18 months	30	35

(b) LIQUID NITROGEN REFRIGERANT

Age	Dairy cattle	Beef cattle
Birth – 1 month	5 seconds	10 seconds
2 – 5 months	7	12
6 – 9 months	10	15
10 – 12 months	12	17
13 – 18 months	15	20
over 18 months	20	25

the same length of time as the iron.

The brand will be visible immediately as a frozen indentation in the skin. As thawing occurs the area will begin to become oedematous and the brands will swell slightly. This will last for up to 48 hours. After this, the area will become dry and scaly. A scab will then form and persist for three or four weeks after which an area of hairless skin will remain. Depending on the stage of the hair growth cycle, the new white hair should grow in within 6 to 10 weeks. If there are too few white hairs to make the brand readable, the irons may not have been cold enough due to water in the coolant or the hair may not have been clipped closely enough. Too short a branding time will also result in poor brands. If the iron is kept on too long the skin at the site will be damaged and result in contraction and narrowing of the site.

When branding white animals, the iron should be kept on the site for 1½ to twice the normal time. This will result in an overbrand that will kill the hair at the site so the resulting mark will resemble a hot brand.

Sheep, goats and small black pigs can also be freeze branded. Sheep are normally branded on the face as the wool cover on the body makes reading brands difficult. Freeze branding of pigs is difficult because of the thick skin and deep-rooted hair which leads to damage to the surrounding tissue during branding. This may lead to scarring with resultant contraction and narrowing of the brand.

2.10 LIVESTOCK RECORDS

The purpose of this section is not to tell you what records you, as a livestockman, should keep but rather to advise you of some of the reasons for keeping records, the types of information you might want to keep and to point out several methods of record keeping.

There are many reasons for keeping records as they are of great value as aids to management and are necessary for financial analyses. On the management side, records are necessary for selecting breeding stock, for knowledge of ages required for vaccination, weaning, marketing, changes in feed and when to breed. Records are also needed to identify animals which produce well, those that have chronic health problems, those females which fail to produce, those males with fertility problems, etc. Financial records are necessary for tax reports, financial statements and evaluating the profitability of the enterprise.

The types of information kept will depend entirely on the use to which the records are to be put. You will find that almost every type of information will be kept on the livestock at an Experimental Station whereas many farmers keep only those records which they can remember. The type of records you keep will depend on the sort of livestock operation you are running. In you have a breeding herd or

flock, you will want information on reproductive history
for each dam and sire as well as information regarding the
productive ability of the offspring. On the other hand,
if your primary business is fattening stock you will be
more interested in rates of growth and feed consumption.
In almost any type of operation you will want to keep a
current inventory of livestock showing purchases, sales,
births, deaths and periodic valuations of the stock.

Selecting the sort of information to be recorded is
critical as too little may result in not being able to
evaluate the enterprise adequately. Too much can result
in no records at all as record keeping then becomes such a
chore that it is dispensed with. It is much better to
have a few accurate records than to have masses of in-
accurately recorded information which is never summarised
or consulted.

Probably the most common method of record keeping is
memory. This method may be sufficient if the memory is
adequate for the number of animals involved, if the person
who carries the records does not suddenly die or go away,
if the memory is accurate and if the memory is long lasting. Permanent written records may be in the form of a
diary with all events relating to the stock entered each
day. Alternatively, a record book may be kept in which a
page is assigned for each animal or litter so information
relative to any animal or litter is readily accessible.
Another possibility is the use of a loose-leaf notebook to
which pages can be added or removed as animals are moved
into or out of the herd. This method has the advantage
that the animal can be sorted according to age, sex or
other convenient classification and changed if the classification should change. It is useful to mimeograph
(cyclostyle) forms for the notebook such as the 'Cow
Record', the 'Ewe Record' and the 'Health Record' in Figs.
23, 24 and 25. The records can be designed on any format
which suits you and includes the information you require.

There are also various types of punch cards available

COW RECORD

NO.	BREED	HERD BOOK NO.	ORIGIN	VALUE IN	DATE OF PURCHASE
	Sire............	{ G. sire............ { G. dam............	Date of birth		
	Dam............	{ G. sire............ { G. dam............	Date of disposal		
			Reasons for disposal		

date of service	bull used	date of calving	sex of calf	no. of calf	wt. of calf	disposal of calf	wt. of cow at calving	Health date	Record treatment

Figure 23. Cow record.

73

EWE RECORD

Number	Breed	Sex	Date of Birth	Twin or Single	Origin	Date of Purchase

Sire.................. { G. sire..................
                        G. dam..................

Dam.................. { G. sire..................
                        G. dam..................

Other Information

Reasons for disposal

Date of lambing	Sex of lambs	Ear numbers	Birth wt	12 Week Weight			18 Week Weight			Disposal		
				Date	Actual	Adjusted	Date	Actual	Adjusted	Live wt	Carcass wt	Age

Ram used

Figure 24. Ewe record.

HEALTH RECORD

	Vaccinations		Worming record	
Date	Vaccine		Date	Remedy
Disease	and wound	treatment		
date	complaint	treatment		

Figure 25. Health record.

which provide for easy sorting by various classifications. 'Keysort' cards are designed with holes around the edges which can be punched out when various operations such as castration, docking, dehorning, etc., have been carried out. In addition, coding can be designed so punches for year of birth, sex, breed, etc., can be made; whenever new information is written on the card, the appropriate holes can be punched. Sorting on any coded category is then accomplished manually by running a needle through the appropriate holes. When the pack of cards is lifted, those that are punched will fall away leaving a sorted deck. These cards can be purchased with a form already printed on them or can be obtained blank so you can type or print on your own form.

The recent development of the personal home computer has made record keeping much easier for the large live-stock operations. With this sort of instrument the oper-ations for each day are stored in the computer memory. With suitable programs the computer will make various production analyses, provide printouts regarding farm, herd or individual production or may be programmed to pro-vide a daily choresheet. This is a listing of all oper-ations such as vaccinations, weaning, drying off, candling etc., that should be carried out that day. The computer will also give warning of sows that are due to farrow or cows to calve. The possibilities for these small compu-ters are almost endless.

The above discussion related almost entirely to perma-nent records. There is another class of records that are constantly changing yet extremely vital. For example, in a dairy herd a cow should be rebred 45 to 90 days after calving and pregnancy-tested three months after that. She should be dried off about two months before the next calf is due and receive extra feed for four to six weeks before calving. With only a few animals this sort of information can be kept in a desk diary or with notations pinned on the notice board. When the herd exceeds 10 or 15 cows, however, these systems become inadequate. One method for handling this sort of information is a circular calendar. This can be made of metal or pin board material onto which magnets or pins can be placed. Each pin is marked with the number of a cow and is moved as each operation such as calving, breeding or drying off occurs. Each day the calendar is turned slightly so in the course of a year it will go through one full revolution. If the pins are properly placed on the calendar the duties required each day will come up opposite the date. A similar sort of calendar is also available for sow operations.

Once you have decided what records you will keep and

the format to be used, it is imperative that you are accurate and thorough in maintaining them. It does not make sense to try to maintain records that are inaccurate or incomplete as they can only lead to misinformation and possible mismanagement.

2.11 AGEING LIVESTOCK

It is frequently desirable or necessary to estimate the age of livestock for which no records are available. The most common method for age determination in cattle and sheep is by means of the teeth. Because the temporary (deciduous) incisors are replaced by permanent incisors at more or less regular time intervals, the number of permanent incisors gives a fair estimate of the age of the animal up to about four years. After four years, estimates are usually based on the amount of wear on the permanent incisors and, as a result, usually are less accurate than the earlier estimates. This is not a serious problem, however, since it is usually more important to know whether a cow is 2 or 4 than it is to know if she is 7 or 9 years old.

Cattle, sheep and most other ruminants have four pairs of permanent incisors on the lower jaw only, which are named as shown below. Pigs and horses have three sets of incisors and one set of canine teeth on both the upper and lower jaws which are named as shown in Fig. 26.

Table 8 shows the dentition pattern for the various

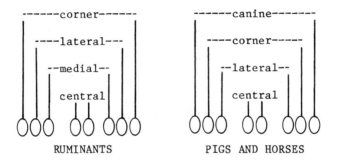

Figure 26. Incisors of ruminants and pigs and horses.

Table 8. Dentition patterns.

COW		
2 years.	oooOOooo	Permanent centrals.
2½ – 3 years.	ooOOOOoo	Permanent medials.
3 – 3½ years.	oOOOOOOo	Permanent laterals.
4 years.	OOOOOOOO	Permanent corners.

SHEEP AND GOAT		
14 – 18 months.	oooOOooo	(Two tooth): permanent centrals.
20 – 24 months.	ooOOOOoo	(Four tooth): permanent medials.
26 – 30 months.	oOOOOOOo	(Six tooth): permanent laterals.
32 – 36 months.	OOOOOOOO	(Full mouth): permanent corners.
4 – 6 years.	OOO△△OOO	'Swallow tail'.
Aged.	OO OOOOO	'Broken mouth'.

HORSE		
2½ years.	ooOOOoo / ooOOoo	Permanent centrals.
3½ years.	oOOOOo / oOOOOo	Permanent laterals.
4½ years.	OOOOOO / OOOOOO	Permanent corners.
5 years.	O OOOOOO O / O OOOOOO O	Permanent canines.

PIG		
7 – 8 months.	oOoooOOo / oOoooOOo	Permanent corners in both jaws.
9 months.	OOoooOOO / OOoooOOO	Permanent canines in both jaws.
12 months.	OOoOOoOO / OOoOOoOO	Permanent centrals in both jaws.
17 months.	OOoOOoOO / OOOOOOOO	Permanent laterals in lower jaw.
18 months.	OOOOOOOO / OOOOOOOO	Permanent laterals in both jaws.

N.B. When ageing a pig by looking into its mouth be extremely careful as pigs can, and do, bite.

livestock species with the small o representing temporary
teeth, the large O permanents, and a blank missing teeth.
Table 8 looks complicated and hard to remember but is

simplified if, for cattle, you remember '2 teeth, 2 years'. Then, simply add 6 months for each additional 2 teeth until the last when you should add a year. To extend this to sheep, determine the age as for cattle then subtract 6 months. It must be emphasised that these figures are approximations as the eruption of teeth, like the age at maturity, is extremely variable.

The first step in ageing cattle and sheep is to look at the animal. The temporary incisors are often badly worn before the permanent centrals erupt. These worn temporaries appear similar to badly worn permanent incisors but the overall appearance of a 2-year old heifer is very different from that of a 12-year old cow.

Some cattlemen maintain that they are able to determine the age of a cow by means of the rings on the horns. Each pregnancy produces a groove at the base of the horns which results in a permanent ring around the horns. Thus, if a cow first calves at two years and produces a calf every year thereafter, her age in years will be the number of rings plus two. The number of rings on the horns is probably better for determining the reproductive history of a cow rather than her age, especially in a situation where calving tends to be irregular or the calving interval long.

2.12 WEIGHT DETERMINATION - ESTIMATES

Livestockmen often need to know how much their animals weigh. Possible reasons for this include management decisions such as how much to feed, when to breed or when to market, or it could be for determining dosages of various medications and vaccines.

Probably the most used, easiest and least accurate method of weight determination (except by highly skilled personnel) is visual estimation. This is a skill developed through practice by estimating the weight of numerous animals and checking your estimate against a weighband (2.13) or weighbridge (2.14).

When estimating weights, try to compare the animal
with something of which you know the weight - yourself, a
bag of grain, a quantity of water. Decide whether the
animal is heavier or lighter than the known weight and by
how much. As you practise and continually compare your
estimates to a weighband or weighbridge, you will find
your estimates become more accurate.

As you estimate weights, bear in mind the species you
are working with. Skill in estimating the weights of *Bos
taurus* frequently leads to gross underestimation of the
weights of *Bos indicus* because of the differences in body
structure. You should also remember that differences in
bone structure can be deceptive when estimating weight.
If, for instance, you are quite skilled at judging the
weight of Friesian cows, you may find you underestimate
the weight of Jerseys because their lighter bone makes
them appear lighter than they actually are. You must also
note the amount of wool or hair cover, as a heavy coat can
make an animal appear to be carrying more muscle than is
actually the case. If you are in doubt as to whether you
are seeing meat or body covering you should catch the animal and handle it to be sure.

A white animal always looks bigger than it actually
is. When estimating the weight of an animal, you should
try to place yourself on the same plane as the animal because, if you are below it, it will look too large and if
you are above it, it will look too small.

Reasonable skill in estimating weight is necessary for
the stockman as it will frequently be necessary to know
weights when a weighbridge is not available or its use is
not practical. As an example, assume you are drenching
100 calves for worms. Dose levels are based on the weight
of the animal but it certainly would not be practical to
weigh each animal before drenching. In fact, the common
practice is to estimate the weight of the animal and
drench on the basis of the estimate. Since an underdose
may be ineffective and an overdose may kill the animal,

reasonable accurate weight estimation is essential. You should use every opportunity to develop this skill. You should note, however, that while most stockmen can become fairly proficient in this skill, it is still essential to make regular reference to the weighbridge. Even with the most experienced a tendency can develop to 'drift off course' with time.

2.13 WEIGHT DETERMINATION - WEIGHBAND

There is a close relationship between the distance around an animal's heartgirth and its body weight. This is a nonlinear relationship expressed by the formula:

$W = aG^b$ where W is the weight; G is the heartgirth; a, b are constants.

The values of the constants depend on the species or breed of animal and the units used in the measurement. Linear relationships of the form $W = a+bG$ may provide accurate estimates over narrow weight ranges but, where all sizes of stock are to be weighed, the nonlinear relationship is more accurate.

From repeated measurements of body dimensions and weights, empirical linear formulas have been developed for determining body weight from measurements of heartgirth and body length as measured from the point of the shoulders to the pin bones (L).

Bennet's formula is most useful for estimating weights of finished steers of the British breeds. The formula is:

$W = 1.04((27.5758 \times G) - 1049.67)$ (W in lb and G in inches)

Shaeffer's formula can also be used for sheep and goats:

$W = \dfrac{L \times G^2}{300}$ (L and G in inches, W in lb)

This formula is reputed to be accurate within 5% in most cases when used with zebu cattle.

An empirical formula for pigs is:

$$W = G^2 \times L$$

Note that L is measured from the poll to the base of the tail, G and L are in inches and W is in pounds. When measuring pigs under 150 pounds, 7 pounds should be added to the weight.

An empirical formula for use with East African short-horn zebu cattle has been developed at Bunda College of Agriculture. This formula:

$$W = 0.0004\ G^{2.678} \qquad \text{(G in centimetres, W in kilo-grammes)}$$

was developed using female breeding stock and finishing steers and probably would not be highly accurate with young stock and bulls. It should be noted that this equation in the logarithmic form appears much less formidable ($\log W = \log 0.0004 + 2.678 \log G$).

Once an empirical formula using only heartgirth has been developed for a class or breed of stock, a tape measure marked in weight units can be manufactured. There are a number of these weighbands on the market. Most have several scales for different species and for different breeds within the species. The weight determined with the weighband will not be as accurate as with a weighbridge but should be within 15% of the actual.

To use a weighband, be sure the animal is standing on all four legs with its head in the normal position. Pass the band around the animal just behind the front legs with the scale you want to read facing upward. Pull the end snugly around the animal and place the end on the scale. The weight can then be read opposite the end of the weigh-band. You should practise using the same tension each time the tape is read so that your estimates are consistent. Do not put your fingers under the tape or twist the tape when reading.

2.14   WEIGHT DETERMINATION - WEIGHBRIDGE

The weighbridge or cattle scale used on any particular
farm will depend on the requirements of the stockman, his
preferences and the money available for the purchase of
the scale. The scale may be portable or permanent but it
must provide for easy handling of the stock, accurate
weighing and ease of maintenance, adjustment and repair.

A cattle scale will normally be a part of the crush
system which will direct the cattle into the scale pen
(see 1.4). The pen itself must be so constructed that the
animals can be easily moved in and out, cannot jump or
climb out and will not slip on the floor. The last is important where the cattle are routinely weighed or driven
through the scale because if an animal slips in the scale
it will often be reluctant to re-enter at a later date.

The principle of operation of a weighbridge is that
the mass of the animal and pens is balanced through a series of levers to a readout device. The levers may be suspended on knife edges, may be hung on steel cables or
bands or mounted on ball or roller bearings. The readout
device may be a common spring balance, a dial scale or a
steelyard. The spring balance is the simplest and least
expensive readout but is subject to spring fatigue and
usually cannot be read, for cattle, more accurately than
to the nearest 5 kg. The dial scale is a complicated device which requires a skilled scale mechanic for adjustment and repair but normally has greater accuracy than the
spring balance. The steelyard is a completely mechanical
device, requiring a poise to be moved along a beam for
reading which means it is slower to use and read than the
other two types. The steelyard is, however, less affected
by cattle movements than the other types.

The accuracy of a scale should be checked periodically
by placing a series of known masses upon it. Errors in
weighing can occur if springs lose their temper, bearings
corrode, cables stretch, or the scale become distorted,
etc. The types and amounts of adjustment possible should

be taken into account when purchasing a scale.

When installing a scale, one should be sure that it is mounted on a level foundation of brick or concrete. In addition, provision should be made for access to the scale mechanism and for water drainage. The latter is especially important if the scale is constructed over a pit. The scale should be so positioned that manure and dirt which build up under the scale can be removed conveniently. This is important because this material can cause corrosion and may affect the accuracy of the scale.

When using a cattle weighbridge remember that, though it is designed for weighing cattle, it is still a fairly delicate instrument which can be damaged by rough usage. After each use the scale pen should be thoroughly cleaned to remove faeces and dirt which may cause corrosion. The readout mechanism should be protected from the weather and, where a locking device is provided, the platform should be locked when not in use. This will help prevent damage to the mechanism.

Pigs and sheep are often weighed in a cage suspended from a spring balance. These units are generally portable so they can be moved from place to place. This allows pigs to be weighed without having to move them from their pens and such a scale can be placed temporarily in a sheep crush for weighing the flock. The same principles of use apply to this sort of scale as to the cattle weighbridge. Calibration is especially important as the change in position of the scale may affect the readings. Care should be taken to see that the scale is firmly placed on a level surface. Small lambs or piglets can be weighed by placing them in a bucket and suspending the bucket from a spring balance.

Poultry can be weighed with a variety of small scales such as the sort used in the kitchen. These generally are not highly accurate so other scales should be used if precise readings are required. The chickens will sit quietly on a scale if they are hypnotised as described in Section 8.2.

It is a common error to expect too much accuracy in weighing livestock. High levels of accuracy are difficult to obtain because the animal may be moving about on or in the scale. Weights are also greatly influenced by gut fill. The weights will be increased if the animal eats or drinks just prior to weighing. On the other hand the weights will be decreased if the animal urinates or defaecates just prior to weighing. The accuracy of weighing will also be affected by the weighbridge or scale used. Accuracy within 1% of the body weight is excellent but is seldom achieved; especially with a large stock. More common levels of accuracy are probably around 5%. This lack of accuracy in weighing does not mean that one should be careless in weighing but it should be recognised that such errors do occur during weighing and care should be taken to minimise them.

2.15 DIPPING

Dipping is the control of external parasites with chemical substances which kill the parasites but have little or no effect on the host animal.

Livestock can be dipped for control of various external parasites such as ticks, lice, mange, fleas, maggots and flies. The frequency of dipping, the substances used and facilities depend on the type of operation and the parasite problem in each specific area. Facilities for dipping are discussed in Section 1.5.

Dipping compounds

Arsenicals.   The arsenic-based dips suffer the disadvantage that they can only be used in plunge dips, can be toxic to the animals and often lose their effectiveness as parasites quickly develop resistance to the arsenic compounds. Because the arsenicals have been widely used for many years, several species are resistant to them in some areas. The advantages of arsenicals are that they are inexpensive, stable, completely soluble in water and

the concentration can be checked by a simple tankside
test.

Chlorinated hydrocarbons.   This group includes such compounds as DDT, BHC, toxaphene, dieldrin, aldrin and chlor-
dane. Although parasites may develop resistance to these
compounds, they are often effective against most ticks as
well as lice and mange. Probably the most serious disadvantage is the long-term effect of these compounds,
which, since they are not easily broken down, may build
up in the tissues and cause death to the animal or, if
eaten, may actually poison the people eating the meat.
For this reason the use of several of these compounds (including DDT and Aldrin) is legally prohibited in various
parts of the world.

Pyrethrums.   The pyrethum compounds are probably the
most effective dipping agents but are seldom used because
they are extremely expensive.

Organophosphates.   The organophosphates are highly
effective, safe in use, and stable but they have a shorter
residual effect than the chlorinated hydrocarbons. Only
one species of tick is known to have developed resistance
to the organophosphates. Since they are expensive they
are often used in mixtures with the chlorinated hydrocarbons. If you wish to use a combination dipwash, the
simplest and safest procedure is to purchase it already
mixed as improper combinations can lead to serious problems including ill-effects on the animals or ineffective-
ness of the ixocides themselves.

Dip concentration

Maintenance of the proper concentration of dip solution is
important. Too high a concentration leads to extra expense and may result in animal deaths, especially with the
arsenicals. Too low a concentration will encourage the
development of resistant strains of parasites. Maintain-

ing the proper concentration means not only using the proper amounts of chemical and water when filling the tank for the first time but also, when adding solution, replenishing losses due to stripping and evaporation. The manufacturer's instructions should be followed carefully in both cases. Replenishment at the beginning of dipping is not a factor when using a spray race since, with a race, it is necessary to mix a new solution each time the stock are dipped. Replenishment may be necessary due to stripping when large numbers of animals (over 500) are dipped.

As the livestock proceed through the dip tank or spray race, quantities of the dip compound and water will adhere to the hair. Since the dip compound is only suspended in the water, the water will drain off leaving some of the chemical behind because it sticks to the hair. This process is known as 'stripping' and results in a lowered concentration of chemical as more and more animals proceed through the dip. You should follow the manufacturer's instructions for replenishing the dip to overcome this problem.

Frequency of dipping

The frequency of dipping depends on the climate, the season of the year, the degree of parasite infestation in the area and the life cycle of the parasites to be controlled. In a hot, humid climate, weekly dipping in summer and fortnightly dipping in winter or even more frequent dipping may be necessary. In drier, cooler climates dipping at greater intervals may provide adequate parasite control. Whatever frequency is used, be sure it is often enough (check with the Veterinary Department), that every animal is dipped at every dipping, that each animal is thoroughly covered and that the proper concentration is used. These are essential to avoid development of resistance among the ticks. For treating a few calves and small stock, a 200 litre drum with one end cut off is convenient. The animal is merely plunged into the drum and well soaked, then removed. Arsenicals should not be used for this purpose.

Dip poisoning

Whatever the dipping facility available and the ixocide used, you should have the antidote available at all times because not only the animals but the personnel working with the animals may be poisoned.

Arsenic poisoning causes animals to appear to have pains in the abdomen, to appear restless, groan when forced to move, breath fast and have diarrhoea. Cattle which are known to have swallowed arsenic should be drenched with 100 gm of sodium thiosulphate (hypo) in one litre of water, repeated every four hours. No drinking water should be allowed until after eight hours when a small amount of water may be given. The animals should be allowed to eat to speed excretion of the arsenic.

There are no specific antidotes available for the chlorinated hydrocarbons and organophosphates. Symptoms of poisoning with these compounds are mostly nervous and the only treatment is to keep the animal quiet, provide sedatives if possible and obtain veterinary assistance.

By far the best method of handling dip poisoning is to avoid it. This can, in a large measure, be done by proper handling of the stock before, during, and after dipping and by correct dip management.

Proper handling includes:

Dip only in the early morning and not in the rain.

Move the cattle to the dip slowly and allow them to water and rest before dipping.

Work the cattle through the dip slowly and quietly.

Do not dip sick animals or cows close to calving. The latter should be hand sprayed.

Avoid factors which prolong drying after dipping.

Do not dip cattle more frequently than recommended or in over-strength solutions.

Proper dip management includes:

Being sure the ixocide is thoroughly mixed and at the proper concentration.

If using a spray race, pressure and volume delivered must be correct and all nozzles functioning.

The vicinity of the dip must be tightly fenced and carefully closed after each dipping. This is especially important where arsenic dips are used. The arsenic compounds have a salty taste which may induce stock to lick fences, etc. leading to arsenic poisoning.

Finally, the dip tank will need cleaning out from time to time and great care must be taken with the disposal of spent dip. The best method of disposal is, where practicable, through a soak-away sump.

2.16 HAND DRESSING FOR TICKS

It sometimes happens that some animals temporarily cannot be put through a spray race or dipping bath or, due to a heavy growth of hair or an improperly adjusted spray race, develop localised clusters of ticks. In either case, some form of hand dressing for tick control is necessary.

The animal which cannot be dipped should be treated either by hand spraying or by pouring buckets of dip over it. These are time-consuming and inefficient methods so should be used only when small numbers of animals are involved. The same chemical concentration as for the spray race or dipping bath should be used. Never use a concentration of dip higher than that recommended by the manufacturers. In order to thoroughly wet the animal you should apply at least 15 litres of dip wash to each animal with a high pressure spray pump.

Localised clusters of ticks often occur in the ears and under the tail. Treatment of these clusters first requires removal of the excess hair; especially in the ears.

The area should then be thoroughly soaked with a standard
dipping material diluted with old motor oil (follow the
hand dressing recommendations on the label) or with a
ready-to-use tick oil or grease. These can be directly
applied to the infested area with a paint brush or similar
tool. Prevention of heavy tick infestation in the ears is
especially important in order to avoid permanent drooping
of the ears as a result of the ticks. When using a spray
race, the effectiveness can be increased by clipping the
hair out of the ears so the spray will more readily penetrate. This is often done as a routine practice at the
start of the season when a heavy tick challenge is expected. Clipping the ears is easily done by confining the
animal in a headgate. The heavy guard hairs can then be
clipped from each ear with two or three cuts with blunt-
ended scissors.

2.17   THE USE OF A FOOT BATH, AND FOOT ROT TREATMENT

To maximise production in domestic livestock, it is absolutely essential that the animals be able to walk without
pain or lameness. Animals with sore feet may be reluctant
or refuse to walk to graze or to water. As a result, milk
production of dairy cattle will decline and meat animals
lose condition.

Lameness may result from a purely physical injury or
may be due to infection with bacteria. In cattle, the
most common invading bacterium is *Bacillus necrophorus*
which causes the condition variously known as foot rot,
foul-of-the-foot, panaritium or necrotic pododermatitis.
Foot rot of sheep is caused by *Bacillus nodosus* in asso-
ciation with the spirochaete *Spirochaeta penartha*. Before
bacterial infestation can occur, the skin must be broken
by physical injury. This may be due to tick bites, step-
ping on sharp objects, walking over stones or may arise
after continued soaking from walking in mud or water which
causes softening and cracking of the feet. Any lame animal should be examined as a matter of routine. If the

foot has a particularly foul, rotten smell, the chances are that the animal has foot rot.

Prevention of foot rot is not difficult but must be regularly practised. Muddy areas in pastures and yards should be drained and walkways for livestock cleared of sharp stones. All metal and other foreign objects must be kept away from areas where livestock may walk. When an outbreak of foot rot threatens, the animals should be treated at least weekly by walking them through a footbath containing 5-10% copper sulphate or 5% formalin solutions. The efficiency of the treatment will be enhanced if the floor of the footbath contains longitudinal ridges. These will tend to spread the hooves as the animals walk through, thus ensuring the treatment reaches all parts of the foot. If a footbath is unavailable, walking the animals through a dry 1:20 mixture of copper sulphate and *slaked* lime each day may be effective. Treatment may also be administered by including drugs in the salt or the feed. The last treatments are really only useful for feedlot cattle or sheep, require care in mixing and feeding and are more expensive than the use of a foot bath.

A foot bath is best constructed as an adjunct to the dip. The bath should be at least 10 m long and contain enough solution to submerge the entire hoof as the animal walks through the bath. The foot bath is often placed within the race immediately preceding the dip tank or spray race. In this position the bath can serve to wash manure and dirt off the animals' feet before entering the dip as well as providing a convenient method for foot rot prevention or treatment.

When an animal is found to have foot rot, the infected hoof should be cleaned and trimmed. If any pus pockets are present these should be opened so that the pus will drain. Treat the hoof with copper sulphate or formalin and carefully watch to see that healing occurs.

As with the other disease conditions, if an animal does not respond readily to treatment, veterinary assistance should be obtained. In fact it is preferable that such assistance be obtained before any treatment is given if this is possible.

2.18   PROCEDURES WITH DISEASED OR DEAD ANIMALS

As a livestockman you will inevitably, at one time or another, have to deal with diseased or dead animals. Symptoms of disease for calves, pigs and poultry are listed elsewhere in this book (Sections 5.4, 7.9 and 8.11, respectively). Symptoms for mature cattle and sheep are not listed, but will become familiar with observation. Observation is an extremely important aspect of livestock husbandry in general and of disease control in particular as keen and accurate observation allows health problems to be detected early when treatment is usually most effective. In fact, it has been said that any fool can tell when an animal is dying, most can tell when an animal is severely ill, but a good stockman can tell when an animal is going to be sick tomorrow. The implication of this is that the stockman knows his animals and observes them regularly and carefully so he knows immediately whenever something is amiss.

This section is included as a guide to what you should do when you do notice a sick or dead animal. It cannot be stressed too strongly, however, that no hard and fast rules can be established as the procedures to follow will depend on the diseases endemic in a particular area, the availability of professional veterinary assistance, the numbers of animals involved and the knowledge and skill of the stockman. In no case should a person untrained in veterinary science presume to diagnose and treat livestock illnesses unless someone with the proper training is definitely unavailable. The reason for this is that many diseases have similar symptoms and often, of two diseases with similar symptoms, one may be relatively innocuous while the other may be extremely serious. Frequently only a diagnostic laboratory can differentiate between the two.

In some areas, however, professional veterinary advice is
unavailable and the stockman will be forced to do the best
he can. It should also be pointed out that a number of
techniques and procedures mentioned or described elsewhere
in this volume are best handled by trained veterinary
staff if at all possible. A note to that effect has been
included in each of those sections.

Most countries have Diseases of Animals Acts in operation which make it obligatory for owners or persons in
charge of animals to report *suspected* cases of certain
diseases to the appropriate authority. Farmers and stockmen must familiarise themselves with these regulations.

Handling sick animals

The following descriptions include the steps which veterinary personnel will follow in cases of disease or death.
They are not intended to be a substitute for such people
but rather a guide to the steps that should be taken if
they are unavailable.

When an animal showing symptoms of disease is noticed,
it should be isolated and immediate steps taken to diagnose the cause of the symptoms. Visual examination of the
body condition, coat and skin, mouth, mucous membranes,
behaviour, nose, eyes, faeces, udder and milk, and urine
should be made and any abnormal conditions noted. The
animal's temperature should be taken and details of the
herd and previous history of the sick animal noted. In
cases where a definite diagnosis cannot be made, samples
should be forwarded to a diagnostic laboratory for further
tests. Material forwarded should include blood and lymph
gland smears and all available information about the
history and symptoms of the animal and the herd. Most
laboratories will provide forms, slides, and containers
for this purpose. In cases where infestation with internal parasites is suspected, it is also useful to send
a sample of faeces collected from the animal's rectum to
be used to assess the numbers and types of parasites

present.

Due to the time required for transmission and testing of samples, veterinary personnel will often prescribe an immediate treatment which may either be directed against the suspected causative agent or a general, broad-spectrum treatment in the hope that such treatment may arrest the disease. Until the animal is cured or the disease diagnosed as non-infectious, the animal should be kept isolated from other stock.

During the time of isolation, supportive therapy should be provided. This includes adequate amounts of fresh, appetising feed, clean, fresh water and comfortable surroundings without draughts.

Handling dead animals

When an animal is found dead, particularly one for which no definite diagnosis of disease has been made, extreme caution must be taken. Improper handling may result in the spread of the infectious organism, e.g. anthrax or blackquarter, to other livestock or to the person doing the examination, e.g. rabies. Diagnostic laboratories in many areas require smears (usually spleen, lymph gland and brain) to be taken from every dead animal. These are used to diagnose the cause of death and assist the compilation of statistics regarding the types of disease prevalent in the area. Submission of smears and organs from dead animals may also be done to assess the cause of death and assist the stockman and/or veterinarian to care for the remaining livestock. The first major symptom of several highly infectious diseases is dead animals, therefore speedy transmission of samples from the carcass is imperative.

If you are fortunate enough to have easy access to a laboratory, it is often most convenient to transmit the entire carcass of small animals such as poultry, sheep, goats or small pigs. Carcasses suspected of being infected with foot-and-mouth disease should never be sent,

however. The types of samples required from large animals will depend on the laboratory but all will need as
much information about the history and symptoms of the
animal as possible. Except in instances where certain
diseases such as blackquarter or anthrax are suspected,
the veterinarian will carry out a post-mortem examination
of the carcass. After making observations about the external appearance of the animal, the examiner will follow
a set procedure of dissection making observations about
the internal organs including the tissues, lymph nodes,
stomach, intestines, spleen, liver, gall bladder, kidneys,
bladder, lungs, heart, trachea, and reproductive organs.
Tissue samples will be collected from some of these, as
required by the laboratory to which the samples are to be
sent.

Disposal of a carcass will depend on the reason for
death. The meat should not be used for human consumption
(unless death has obviously been due to attack by a pred-
ator such as a lion or hyena) or unless permission is
granted by the attending veterinary assistant or officer.
In all other cases, the carcass should be deeply buried
after being covered with lime or should be thoroughly
burned. Neither procedure is easy but burial probably is
the easiest. It requires a very large quantity of fuel
and a long time to completely burn the carcass of a
mature cow.

In conclusion, it must be reiterated that the procedures followed in cases of diseased or dead animals are
best carried out by personnel specifically trained to do
so. If such personnel are not available in your area, it
is advisable that, as soon as possible after your arrival,
you contact the veterinary department for information regarding the problems you are likely to encounter and the
procedures you should follow. It is only with such advice
that you will be able to prevent deaths among your livestock and avoid a potentially disastrous spread of disease.

2.19 WOUND TREATMENT

A wound is an injury in which the skin or other membrane
is broken. It is inevitable that livestock will at some
time suffer wounds. Wounds may be either deep or superficial but the objectives in treating any wound are the
same: to promote rapid healing while preventing secondary
infections. The basic methods used in meeting these objectives are:

Cleaning the wound thoroughly.

Preventing the accumulation of fluid in the wound.

Allowing the wound to rest.

Any wound, whether superficial or deep, should be
thoroughly cleaned to remove all dirt, manure, straw, etc.
and treating with an antiseptic to destroy any contaminating bacteria. A wound can be cleaned by swabbing with
a bit of cotton wool soaked in antiseptic solution or by
using a syringe to flush the wound with antiseptic. Any
good proprietary antiseptic designed for wound treatment
can be used but it is important to use the proper dilution
as shown on the bottle. The use of too strong a solution
will delay healing. If no suitable antiseptic is available, 1 teaspoonful of table salt in a half litre of
boiled water works well for washing wounds. Applying this
solution will sting the animal so it must be well restrained. If there is severe bleeding, place a clean cloth
pad over the wound and apply firm pressure until bleeding
stops. When the bleeding has stopped, fix the pad in
position and obtain veterinary assistance. Do not remove
the pad since bleeding may restart. Cotton wool should
never be used directly on a wound as the fibres will stick
to it and may cause contamination.

Draining the wound is very important, so much so that
it may be necessary to cut a drainage path if the wound is
such that it will not drain properly. After washing and
providing for drainage the wound should be treated with an

antibiotic powder or cream. The wound should then be ban-
daged to prevent the ingress of dirt or other foreign
material. Do not pack a wound with salve, ointment or
grease if it is possible to avoid doing so, or they will
keep air from the surface and delay healing.

'Resting' a wound means just that. Avoid unnecessary
washing or probing of the wound. If a clean scab forms
without pus underneath the wound should be left alone and
allowed to heal.

If an infection should occur, it may be necessary to
re-clean the wound and reapply the antibiotic. The wound
should then be re-bandaged and left for several days.
Once a clean scab has formed, the bandage may be removed
and the wound allowed to heal while exposed to the air.

With serious wounds which require stitching, are in
vital places or become infected, it is advisable to get
veterinary assistance.

2.20   FOREIGN BODIES IN THE DIGESTIVE TRACT ('HARDWARE
       DISEASE')

When grazing, cattle pull small bunches of forage into
their mouths with a sweeping action of the tongue. These
bunches are then bitten off between the lower incisors and
upper dental pad and swallowed with very little further
chewing. Due to the rapid rate at which biting and chewing occur when grazing (over 90 per minute), stones or
bits of metal are often taken into the mouth and swallowed
along with the forage. Because of the structure of the
ruminant stomach, most of this material falls into the
reticulum and remains there. Round or smooth objects
cause no problems but sharp materials may puncture the
reticulum or rumen or even the diaphragm or heart. Such
punctures result in conditions known as hardware disease,
traumatic reticuloperitonitis, traumatic gastritis, or
traumatic pericarditis.

The onset of hardware disease is usually sudden with
the animal going off feed and milk production falling

sharply. The animal stands with its back arched and shows a reluctance to move. If it is pinched or pushed on the back it will usually grunt.

Prevention is much easier than cure. Care should be taken to minimise the exposure of cattle to areas where nails, screws, staples, bits of wire, etc. may be ingested. This also means that all such materials must be cleaned up when building fences or carrying out other construction in areas to which livestock will have access. A good stock-man should develop the habit of picking up and properly disposing of any nails, screws or bits of wire found lying on the ground anywhere as these, apart from the danger of hardware disease, may also lead to foot injuries (see 2.17) or flat tyres on farm or other vehicles.

A further precaution against hardware disease involves inserting a bar magnet about 50-75 mm long and 10-25 mm in diameter into the reticulum of each animal. The magnets will hold ingested metal objects so that they have less chance of puncturing the reticulum. This is usually only done with valuable beef animals and dairy cows.

In rare cases, hardware disease may respond to anti-biotic therapy but it is usually necessary to remove the offending objects surgically. The diagnosis of hardware disease and the decision as to how it should be treated should be left to a professional veterinarian, bearing in mind that although an operation is usually successful, it is also time-consuming for the vet and therefore expensive.

2.21 BLOAT

Bloat, also known as hoven or tympany, is an accumulation of gas in the rumen and reticulum. Large amounts of gas are normally produced in these chambers and usually ex-pelled by belching (eructation). If, for some reason, the animal cannot or will not belch, the gas accumulates and first causes discomfort. As the condition becomes more severe, breathing is laboured and the normally depressed area just in front of the left hook will be seen to bulge.

Further distension may cause the animal to go off its feet and it may die.

Causes of bloat

Bloat may be caused by a physical obstruction in the oesophagus such as an apple or potato which prevents eructation or may be due to the gas becoming entrapped in bubbles in the rumen so that eructation becomes impossible (frothy bloat). The exact reason for the formation of frothy bloat is not known but is associated with grazing on lush pastures, especially legumes. Bloat problems normally occur when the animals are first turned into such pastures or where they have access to the pastures only intermittently. Stock which regularly graze lush pastures seldom have bloat problems although there may be individuals which bloat despite continual grazing. If stock are to be turned into lush pastures for the first time, they should be well fed on hay beforehand so they will not overeat in the pasture.

Treatment of bloat

If an animal is seen to be bloated, the oesophagus should be palpated to see if mechanical obstruction is present. If such an obstruction is found obtain veterinary assistance without delay. Moving an obstruction within the oesophagus is a job which requires care and skill as the oesophagus is a tender organ and is easily ruptured. Rupture of the oesophagus almost certainly will lead to death.

If no obstruction is found the animal may be treated with a defoaming agent. The most effective and most easily obtained agents are the vegetable oils such as peanut (groundnut), maize (corn) or soybean oils. Approximately one half litre should be administered as a drench. If the oils are unavailable, cream will sometimes provide relief.

If drenching with the oil is ineffective, a tube can be inserted into the rumen by way of the mouth. The tube

should be well lubricated before insertion. When passing
the tube through the mouth ensure that it enters the oesophagus and not the trachea. This can be checked by feel-
ing the oesophagus as the end of the tube can be felt
through the wall. When the tube is in the rumen, the end
should be moved about so it will enter gas pockets and to
keep rumen contents from plugging the opening.

In severe cases where the animal is prostrate it may
be necessary to puncture the rumen to allow the gas to escape. This is a serious operation as there is a high risk
of infection due to the rumen contents which come into
contact with the peritoneum. If puncture is absolutely
necessary an instrument known as a trocar and cannula
should be used. This consists of a sharp puncturing tool
(the trocar) which is surrounded by a hollow tube (the
cannula). To use this instrument, measure one hand-width
forward from the left hip bone. The swelling will generally be greatest at this point. Using the trocar with the
cannula over it, jab directly into the rumen at right
angles to the swollen flank, never upwards. Hold the
cannula in place in the rumen and remove the trocar. This
should cause the immediate release of the ruminal pressure.
The cannula should be left in place until the bloat is
completely relieved. Some cannulae have small eyes for
tying in place with sutures through the skin. Veterinary
advice should be obtained before the cannula is removed as
this will leave a hole in the rumen which may allow the
escape of the rumen contents into the peritoneal cavity.

2.22   GESTATION LENGTHS

Gestation is the period from service by the male to the
birth of the young. The time varies between species and
between animals within a species. The figures given in
Table 9 are average and several days' variation either way
is not abnormal.

Once a cow, for instance, has been diagnosed as in
calf it is normal practice to compute the date on which it

Table 9. Gestation lengths.

Species	Days	Months
Cattle	284	9½ months
Goats	150	5 months
Rabbits	30	1 month
Sheep	150	5 months
Swine	114	3 months, 3 weeks, 3 days

can be expected to calve. The most common method for doing this is using a gestation table. These are available in many books dealing with livestock husbandry. To use such a table, one enters the table at the date of service and reads across to the date due.

Another method, which looks complicated at first glance but becomes simple with use is as follows:

(a) Write down the service date e.g. 13-2-81 (13 February 1981).

(b) To the day add 7, to the month 9. Our example thus becomes 20-11-81 which is the date the cow is due to calve.

(c) Suppose the service occurs in April or later in the year e.g. 13-7-81 (13 July 1981). Following the same step as in (b) above we get 20-16-81. Since there are only 12 months in a year subtract 12 from the month figure date and add 1 to the year giving 20-4-82 which is the due date.

(d) Similarly if the days column goes over the number of days in the month, subtract the number of days in the month of service from the day column and add 1 to the month. For example, if service occurred on 29-1-81 (29 January 1981) the cow is due 36-10-81 or, since there are 31 days in January, 5-11-81 (5 November 1981). Several examples are shown on page 102.

With use and practice, it is possible to use this method without recourse to pencil and paper. This will avoid the necessity of always having to find a gestation table.

Service date	Intermediate	Date due
1-1-81		8-10-81
25-1-81	32-10-81	1-11-81
4-5-81	11-14-81	11-2-82
27-8-81	34-17-81 (3-18-81)	3-6-82

Computation of expected kidding or lambing dates for goats or sheep merely requires addition of 5 to the month of service. If this gives a total over 12, subtract 12 from the month figure and add 1 to the year, For example:

Service date	Intermediate	Date due
1-6-81		1-11-81
13-11-82	13-16-82	13-4-83

For pigs, add 24 to the day and 3 to the month then proceed as described for cattle.

Service date	Intermediate	Date due
6-5-83		30-8-83
10-8-83	34-11-83	3-12-83
14-12-81	38-15-81 (7-16-81)	7-4-82

3   Cattle in General

3.1   BEEF TYPE

The beef animal is a meat factory, either producing a carcass for consumption or producing offspring which will
ultimately be consumed. The ideal characteristics required of a beef animal depend on the demands of the market in which it is to be sold and on the type of feeding
it will encounter.

Beef production can be divided into two basic systems:
intensive and extensive. Intensive beef production en-
tails keeping the animals in confinement and carrying high
quality feed to them so that they grow and fatten rapidly
to market weight. Extensive beef production, on the other
hand, involves allowing the animals to graze large areas
of cheap pasture and range from which they derive the nutrients for growth and fattening. A combination of the
two is probably most common, so that the animals are kept
at low cost under extensive conditions until they reach a
size of frame (i.e. skeletal size) at which they will
thicken their muscles and fatten efficiently. They are
then placed in stalls or feedlots and fed a diet containing a large proportion of concentrates and high quality
roughage until they reach market weight with thick muscles
and some fat. At this stage they are said to be 'finished'.

The aim of beef production should be to provide an
animal which carries a high proportion of meat in the body

areas (the hind quarters and loin) which are worth most because of the thickness of muscles, and will grow rapidly and efficiently on the type of feed available.

The beef animal derived from the European breeds (Fig. 27) is a short 'blocky' animal with short legs, and a short neck. The head is more square than that of the dairy cow but is not coarse. This type of animal will grow rapidly on concentrate feeds and good pasture and will 'finish out' at 400 to 600 kg liveweight and yield a high grade carcass weighing 200 to 300 kg at 18 months of age. Generally these cattle will require continuous intensive feeding on concentrates to reach desirable market condition. The beef animals derived from zebu cattle (Fig. 28) are generally more upstanding and have a less blocky appearance than the European sort. They are characterised by a hump on the shoulders or the neck. This type of animal grows well on pasture but will generally not reach slaughter weight before three years of age. The size of the resulting carcass varies widely depending on the location, feed and breeding, but it usually will not be as heavy as the carcass from the European sort of the same age. In addition there will be less fat, particularly on the outside of the carcass. Zebu type cattle will respond to intensive feeding but usually not as

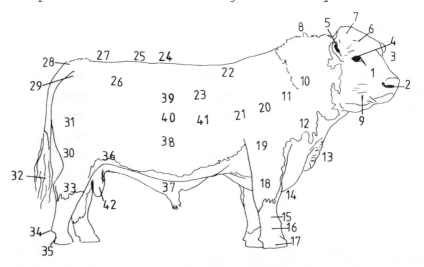

Figure 27. Parts of a beef animal of European breeding.

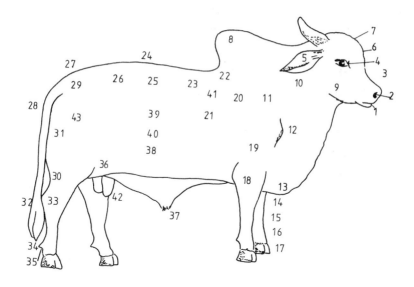

Figure 28. Parts of a zebu beef animal.

efficiently as the European types and will not lay down as much fat. Crosses between the Zebu and European types are frequently used as they combine the hardiness of the former with the fattening ability of the latter. Some of these crosses have been regularly made and the resulting offspring are considered breeds in their own right. Table 10 shows some common breeds of the three sorts.

Table 10.

European	Zebu	Crossbreds
Aberdeen Angus	Afrikander	Beefmaster
Charollais	Angoni	Bonsmara
Chianina	Boran	Brangus
Criollo	Brahman	Charbray
Hereford	Butana	Jamaica Red
Limousin		Santa Gertrudis
Lincoln Red		
Red Poll		
South Devon		
Sussex		

Beef parts.   (see Figs. 27 and 28)

1. Muzzle; 2. Nostril; 3. Face; 4. Eye; 5. Ear; 6. Forehead; 7. Poll; 8. Crest (Hump); 9. Jaw; 10. Neck;
11. Shoulder vein; 12. Point of shoulder; 13. Dewlap;
14. Brisket; 15. Knee; 16. Shank; 17. Foot; 18. Forearm;
19. Arm; 20. Shoulder; 21. Heart girth; 22. Top of shoulder; 23. Crop; 24. Back; 25. Loin; 26. Hook; 27. Rump;
28. Tail head; 29. Pin; 30. Twist; 31. Quarter; 32. Switch;
33. Hock; 34. Dewclaw; 35. Pastern; 36. Flank; 37. Sheath;
38. Paunch or Middle; 39. Loin edge; 40. Rib; 41. Forerib;
42. Scrotum (on bull), Cod (on steer); 43. Thurl.

Although modern production in developed countries usually requires separate herds for milk- and meat-producing stock the most common cattle in the rest of the world are used for both meat and milk and may be used for other purposes, such as draught, as well. These stock will not produce milk as efficiently or in such large quantities as dairy cattle and their meat production will also probably be somewhat less than for beef cattle but, because they perform a dual role in production, the decrease in each of the forms of production is tolerable. Some breeds of multi-purpose *Bos taurus* cattle include:

Fleckvieh	Gelbiveh	Meuse-Rhine-Yssel (MRY)
Friesian	Dairy Shorthorn	Brown Swiss

3.2   DRIVING AND MOVING CATTLE

Cattle are large, strong, have minds of their own, and can be difficult to work with or even dangerous if improperly handled. This section deals with some techniques of working with cattle which allow the smaller, weaker human to get larger, stronger cattle to do what the human wants without a contest of strength which the human will probably lose.
    Excited cattle are difficult to handle because their movements are unpredictable and more rapid than normal.

To avoid getting them excited, they should be handled
quietly without shouting, loud whistling or beating with
sticks. They should be talked to quietly and steadily in
a soothing tone of voice. The personnel around the cattle
must use smooth, definite movements and not be afraid of
the cattle. Respect their size and strength but do not
fear them as fear is contagious and will upset the stock.

Cattle being driven in a herd should move at their own
pace. A herd moving at the proper pace will be in the
form of a triangle with the boss or dominant cow in the
lead. If the drovers push them faster than their normal
pace, the rearward cattle will push up on the forward animals thus flattening the triangle. If the cattle are
pushed much too fast, the herd will split and travel in
two groups at right angles to the desired direction. The
diagrams in Fig. 29 illustrate this. In each case the
desired direction of travel is to the right. The arrows
indicate the actual direction of travel.

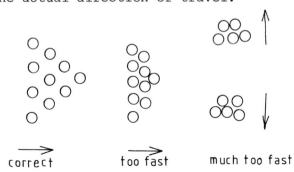

correct   too fast   much too fast

Figure 29. Effect of driving speed on the shape of a cow herd.

If an animal breaks away from the herd, do not run after
it to bring it back. This only serves to drive it further
away. Instead, walk at right angles to the breakaway's
direction of travel and gradually curve around until it is
between you and the rest of the herd. Then, talking continually, work it slowly back to the herd.

If an animal comes at you, never turn around and run
away from it. Keep your eyes on it and sidestep out of
its way or, if necessary, climb the fence. If the latter

is necessary, you must continue to watch the animal.

It is often difficult to drive animals singly as they do not like to leave the rest of the herd. The best way to handle this is to drive two or three other animals with the desired one to the destination, then return the others to the herd. Be sure to take at least two extra so you do not have the problem of a single animal on the return trip.

Driving a cow with a calf or, even worse, driving a cow away from her calf, can present special problems. These will depend on the situation, but generally, if driving both, you must travel at the calf's speed. If you drive the cow ahead of the calf she will attempt to circle back to it which will create disturbance so it is best just to move at the speed of the calf in the first place as it will be nearly impossible to move faster anyway. Driving a cow away from her calf can be difficult, especially with zebus and other cattle which tend to have strong mothering instincts. Do not do it unless necessary but, if it is, be calm but firm. Do not allow the cow or calf (or yourself or your helpers) to become excited. It is often useful to get the calf out of the cow's sight, e.g. into a shed. If the calf is moved so the cow does not know where it has gone, it will often be possible to drive her away from the area.

When working with cows and calves, particularly if the cows are nervous about the calves, watch for the cow which may turn on you. This normally is to protect her calf and as soon as you move away from the calf the charge will stop. You will occasionally find a cow which just does not like people and will charge at every opportunity. The only way to handle this is to avoid presenting opportunities and to be ready to move quickly should a charge occur. Such an animal should be disposed of as soon as possible.

Animals that are handled frequently are often trained so they can be led rather than driven. This may be by means of a halter (see 1.8), a rope around the horns or a nose ring. For stock which will be moved frequently (e.g. oxen, bulls or cattle for shows) it is well worth the effort to teach them to be led on a halter. If possible, this should be done while they are young when they are easier to handle and tend to train more easily. If bulls are to be led they should have a ring installed in the nose. Leading will be done with a halter but a bull staff (a 2 m pole) should be hooked into the ring to control the bull if problems arise.

When working cattle through a crush, it is especially important to keep the animals calm. Some animals will be reluctant to move into the crush but can often be convinced by a larger animal pushing from behind (only the next one, do not try to move a reluctant animal by moving another four or five behind it), a slap on the rump with the flat of the hand or having its tail twisted. Do not use sticks, they are no more effective than the hand and cause bruising. With especially difficult cattle an electric stock prod can prove useful. Care must be taken not to over-use the prod as this tool tends to excite the animals and increase the difficulty in handling. In an *emergency* situation where no prod is available, biting the animal's tail is effective. To do this, grasp the tail near the switch and bite hard. A gentle bite is useless and, once you overcome the disinclination to put the tail in your mouth it matters little whether you bite gently or not, so you might as well make it effective. If the animal does not respond, turn the tail slightly and bite again. The prod or bite are also useful with the animal which 'goes down' in the crush and refuses to get to its feet.

Make sure no one is standing along the crush or in such a position that they appear to the cattle to be blocking the crush. This simple but frequently overlooked error can lead to chaos with an otherwise manageable bunch of cattle.

When you wish to pack a crush with all heads to one

side, e.g. for drenching, you can make the animals move their rear quarters in the direction you desire by twisting their tails in the proper direction. To move an animal to the right, grasp the tail and twist it over the back from the left side. A crush can be packed quickly and smoothly with a man on each side of the crush. The man on the 'head' side passes each animal's tail over its back to the man on the tail side who pulls the animal over and holds it until the next moves in and the operation is repeated. Do not use excessive force or the tail may be broken.

In summary, work cattle quietly and at their own pace; do not excite them or fear them, work them closely and be sure and definite in your movements. It is very easy to tell how good a cowman is by how he works his cattle. The one who works a bunch easily without undue exertion or excitement is probably not blessed with easy animals, it is his skill that makes them easy to handle.

3.3   RESTRAINING CATTLE

It is often necessary to confine cattle heads for drenching, checking ear tags or tattoos, taking blood samples, etc. This can be done with a halter, rope on the horns or nose ring, but these are slow and not worthwhile when large numbers of animals are being dealt with. As instrument known as a bulldog or bull leader can be used. This is merely a clamp inserted into the nostrils. The fingers can also be used in the nostrils but they rapidly tire. The method preferred by the author is to catch the animal in a crush, then stand with the right hip against the left side of the animal's head. The thumb of the left hand is placed into the animal's mouth between the incisors and molars with the fingers under the jaw. The right hand is used to grasp the right ear or horn. This gives a very secure hold on all but the largest animals. Holding the horns alone is unsatisfactory because you are only pitting your strength against the superior strength of the animal's

neck. Also, grasping the horns causes it to shake its
head. When catching an animal to hold it by the nostrils
or jaw, do not try to grasp a horn first as the head shak-
ing will prevent you from further securing the animal.
Also, it tends to excite the animal. If the animal holds
its head down while you are trying to catch it, you can
lift the head by standing nearly in front of the animal
and lifting on the lower jaw. Once the head is lifted the
lower jaw can be held with the thumb in the mouth.

To lift the front foot of a cow, tie its head so it
cannot move about then stand beside the animal facing
backward, reach down and lift the foot, bending it at the
knee. Be ready; the animal will probably lean its weight
on you as the foot comes off the ground. Lifting the hind
leg is difficult and should not be tried without a proper
stock to confine and support the animal. To cause an animal to lift its foot a short distance, e.g. to get its
foot off a rope you are working with, exert upward pressure with your foot on the underside of the dew claw of
the appropriate foot. This usually will cause the animal
to lift its foot clear of the ground and replace it. Frequently the foot will be returned to its former position
so you must be on the alert and free your rope, etc. while
the foot is in the air.

Cattle kick! There is no doubt that cattle can, and
do, kick and the frequency increases when they are being
treated in a manner to which they are unaccustomed. There
are several ways of preventing kicking. Often tying the
animal's head high so its neck is stretched upward will
prevent kicking. Another simple method is to raise the
tail straight up over the back. A rope passed tightly
around the animal just in front of the hips (udder kinch)
will also stop kicking. Another method is to tie the legs
together with a figure-of-eight loop around both hind legs
just above the hocks or to tie a loop tightly on one leg
so as to put pressure on the tendon above the hock. All
the above should only be used with cattle which are

handled infrequently. When frequent handling is necessary, e.g. for milking, the animals should be trained so restraint is unnecessary.

Whenever working cattle (except for driving them), it is imperative that you work close to the animals. If there is danger of kicking, it will be a less severe kick if it travels 75 mm than if it travels 500 mm. Standing back hoping to dodge the kick is useless as you will not have time to respond once the foot is in motion. When working at the head, have your body in contact with the animal. This gives you a more secure hold and you often can sense a movement before it begins. This is the same sort of situation as the athlete who watches for the muscular signals of what his opponent is going to do, not what he is doing. Since most cattle do not 'fake' as athletes do, the tensing of muscles or small movements often allow you to take corrective action before the movement becomes uncontrollable. You will not sense these movements if you are not in close contact with the animal.

3.4 CASTING CATTLE

It is sometimes necessary to place a cow or other large animal on the ground for examination or treatment. Mobbing the animal and pulling it down or wrapping its legs in ropes usually succeeds in that the animal ends up on the ground but the animals or the people stand a good chance of being injured.

Figure 30 shows a method of casting cattle (the Reuff Method) which minimises the chances of injuring either the animal or the people. The loop on the neck should be tied with a bowline or other non-slip knot. The rope is then passed around the heart girth, passed back and looped around the animal just anterior to the hip bones. A steady pull on the free end will cause the animal to collapse slowly. It is advisable to cast the animal on soft grass or a bed of hay or straw.

An alternative method is shown in Fig. 31. In this

Figure 30.   Reuff method of casting.

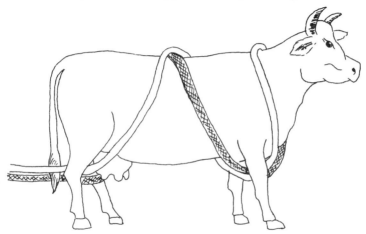

Figure 31.   Crisscross method of casting.

method the middle of the rope is placed over the neck and
the ends passed between the front legs. They cross under
the brisket and are then passed upward and crossed over
the back and then downwards past the flanks and between
the hind legs. Traction on the free ends will then cause
the animal to collapse. The ropes *must* cross under the
brisket (not on the throat) or the animal will not go down.

It is not known exactly why either method works but it
is interesting that some animals respond better to one
method than the other but neither seems to be the 'best'
method. One great advantage of the latter method is that,

because there are no knots involved, it is easier to remove the rope when you are finished with the animal. With
the Reuff method the rope tends to stay in position better.
This is an advantage when casting wild or excited cattle
as the rope can be put on in the crush. The animal is
then let out of the crush and cast.

With both methods it is necessary to confine the animal so it cannot walk backwards when the rope is pulled.
Once the animal is down, it can be kept down by maintaining tension on the rope.

Whenever using ropes with livestock, it is very important that you do not become entangled in the rope otherwise you may be dragged and severely injured if the animal
should bolt. Not becoming entangled means never wrapping
the rope around your hand to pull it. You also should be
careful not to step on the rope or into loops of the rope.
The latter is made easier if the rope is kept neatly
coiled for as much of the time as possible.

Calves can also be cast using a rope but this is a
rather slow and inefficient method for handling small animals since they can be easily cast by hand without ropes.
The diagrams in Fig. 32 show two methods of grasping the
calf to accomplish this. The method shown on the left is
adequate if the operator is fairly tall and the calf small
whereas the method shown on the right can be used for any
calf providing the flanker can lift its feet off the
ground.

The procedure as follows is with the flanker on the
right side of the calf. If the flanker is on the left,
the sides are reversed.

Stand beside the calf. Place your right hand over the
neck and under the throat on the calf's left side.
Grasp the left flank with your left hand.

Lift the calf off the ground.

Push the legs away from you using your left knee.

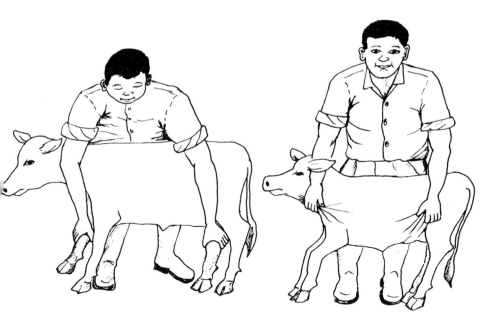

Figure 32. Flanking calves.

Allow the calf to slide down your left leg to the ground.

As the calf goes down, allow your right knee to bend so that as the calf reaches the ground you can place your right leg across its neck.

After the calf is on the ground readjust your right leg so that it is fully across the neck immediately behind the ear. Put your full weight on your right leg. You will not hurt or strangle the calf doing this as the thick muscles of the neck will protect the windpipe.

After the calf is down you may tie it as shown in Figs. 33 and 34. This method of tying is satisfactory for small calves and for minor operations (e.g. taking temperatures) but for larger calves and more major operations, it will be necessary to use a more secure tie.

A calf can be more securely held without tying if the person on the neck grasps the upper front leg at the pastern and doubles the leg at the knee and holds the shank tightly to the forearm.

Figure 33. Placing the ropes for tying a calf.

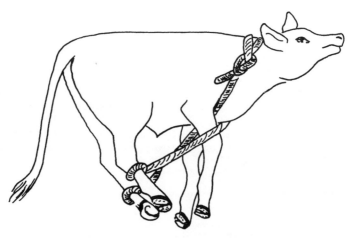

Figure 34. The calf after tying.

3.5 SLAUGHTER

It is often desirable to slaughter an animal on the farm for home consumption or private sale of the carcass. The actual technique of slaughtering and dressing a beast is almost impossible to learn from a book and is a procedure which must be learned by demonstration. The techniques also vary for the various species, from country to country, or even between areas within a country so the complete description of all methods could (and does) comprise a book in itself. Thus, the purpose of this section is not to describe the various slaughter techniques, but to point out a number of general principles which apply to any slaughter operation.

1. *The animal to be slaughtered must not be allowed to become excited* as this may lead to meat which has a higher pH, darker colour and is drier than normal meat. If, for some reason, the animal does become excited, it is best to postpone the slaughtering or use another animal.

2. *Be humane. Do not allow the animal to suffer.* It is normally advisable to stun the animal with a hammer or gun prior to slitting its throat. If this is not possible, cutting the spinal cord with a sharp knife in the hands of a skilled person is quite effective and fast.

3. *As soon as the animal is stunned, the throat should be slit to allow complete bleeding.* If possible, the carcass should be hung with the head downward. If this is not possible placing the carcass with the head downhill will promote thorough bleeding. This is desirable since excess blood in the muscle will darken the meat and reduce its keeping quality.

4. *The purpose of slaughtering is to produce meat and it is imperative that the meat be as clean as possible.* Great care should be taken to avoid contamination by flies, faeces, dirt, and intestinal contents. If you are

slaughtering without the aid of a hoist, care should be taken that the carcass remains on the hide side at all times and is not allowed to touch the ground. When removing the intestines be careful not to puncture them and spill the contents inside the carcass.

5. *At all stages, your work will be easier and more efficient if only very sharp knives are used.* All knives should be sharpened before slaughter commences and periodically honed with a steel during the process. After slaughter is completed, the knives should be thoroughly washed in hot, soapy water, rinsed and dried. When completely dry they should be stored in such a manner that the blades will not come in contact with metal or concrete so they will not become dull.

6. Two cardinal rules of the abattoir should be:
(a) *Use knives to cut meat and saws (or cleavers) to cut bone.*
(b) *If you drop a knife, let it fall.* Do not attempt to catch a falling knife as this may result in an extremely severe wound on your hand.

7. *After the carcass has been skinned and gutted, it should be reduced to easily manageable sections and placed under refrigeration.* The latter is important since quick cooling will slow bacterial action and consequently will slow the spoilage of the meat.

8. *The area used for slaughtering should be thoroughly cleaned* of blood, bits of tissue and bone and other debris after the carcass has been removed. This will prevent fly infestation and keep the area clean for further slaughter operations.

3.6 DEHORNING

Apart from aesthetics, horns have few useful purposes for domestic cattle. In some cases they may be useful for

defence against predators, for controlling work oxen or
for attaching a yoke. On the other hand, there are several disadvantages with horned cattle. Animals with horns
tend to fight more than those without horns and cause more
damage to hides and meat when they do fight. In addition,
such fighting frequently results in open wounds which can
be further complicated by infection or fly strike. Horned
cattle which are fed in troughs or bunks require as much
as three times as much space per animal as dehorned or
naturally polled cattle. Horned animals are also more
dangerous to the people working with them.

It is desirable to remove the horns as early as possible because of the problems of restraining the larger
animals and the greater setback suffered if removal is de-
ferred. It sometimes happens however that calves are not
disbudded and it is necessary to remove the horns at a
later age. The only method for totally removing horns
from large stock creates open wounds which will bleed
severely and are liable to infection. There are commercially available instruments known as guillotines which
may be used to cut off the horns. The horns can also be
removed with a saw, embryotomy wire or a length of wire
rope about 3 mm in diameter such as brake cable from an
automobile. To make this cut, one or two strands should
be removed along the entire length to be used for cutting.

Preparation

Whichever method is used, the animal will need to be cast
or closely restrained in a bale. Closely clip the hair
all around the base of the horn so you can see where the
soft tissue ends and the horn proper begins. Locate the
groove in the bone that runs from the corner of the eye
socket to the base of the horn. At about one-third the
distance to the eye from the horn inject 6 ml of local
anaesthetic subcutaneously. Be sure to check that you are
not in a blood vessel as injection into the circulatory
system will carry the anaesthetic away from the horn.

Inject about 3 ml in the direction of the base of the horn
and 1.5 ml each to the left and right. Allow at least 5
minutes for the anaesthetic to take effect. The use of an
anaesthetic for such an operation is compulsory in many
countries.

Removal of the horn

When using the guillotine, open the handles as far as they
will go and slip the horn between the blades. Close the
handles until the blades just touch the horn base to be
sure that you are in the correct position. This should be
just at the junction of the skin and the base of the horn
so that all of the horn is removed all the way round.
Once the guillotine is in the proper position, close the
handles all the way in one motion. With young stock where
the horns are not yet attached to the skull a scoop or
Keystone dehorner can be used in a similar fashion.

Use of an ordinary wood saw requires that the head be
held securely in position while the cutting takes place.
Before starting the cut, be sure that the saw is positioned so the entire horn will be removed.

Embryotomy wire is made for dismembering dead foetuses
within the uterus. It has rough burs on it so if handles
are attached to two ends of the wire and the wire drawn
back and forth in a sawing motion the wire will cut
through bone and tissue. To dehorn with embryotomy wire
use a piece about 1 m long with handles on both ends. Get
into a comfortable position as the cutting may take a few
minutes and once you start you should not stop. Be sure
the wire is correctly located to remove the entire horn
and saw it off using a back and forth motion on the two
handles. Dehorning with a piece of wire rope is very
similar to the use of embryotomy wire but slower, although
bleeding is said to be less with the wire rope. Because
of the slower cut, more heat is generated during the cutting so the cutter tends to cauterise the wound. When the
horns to be removed are 60 to 150 mm long an elastrator

band may be used. This is a strong rubber ring which may also be used for castrating and docking. To dehorn with the elastrator merely place the band at the base of the horn which will cut off the blood supply. After four to eight weeks the horn will drop off. The animals should be carefully observed for development of infection or fly strike.

Post-operative care

The horn has a profuse blood supply so whichever method you use, several arteries will be cut. This may result in copious bleeding, especially with the guillotine or scoop where the arteries are cut smoothly. Bleeding is less when the saws are used as the ends of the arteries are ragged and tend to stop the flow. The bleeding will stop by itself but, if you prefer, steps may be taken to stop it. Use of forceps to grasp the ends of the arteries and pull them off will make ragged ends. The wound may also be packed with cotton wool and painted with Stockholm Tar or antiseptic or an astringent may be used. Cauterisation of the wound with a hot iron such as a disbudding iron may also reduce bleeding although it is very difficult to stop completely using this method.

With mature or nearly mature cattle the frontal sinus extends up into the base of the horn and when the horn is cut off a hole into the sinus will be left. This is not serious and will quickly close itself if infection is prevented. When dehorning at any time but especially during the fly season it is important that this opening be checked every day or so to ensure that infection is not developing. If there is infection, thoroughly flush the sinus with a mild salt solution (0.9% is best) and pack the cavity with antibiotic powder such as sulphanilamide. The wound should then be repainted with Stockholm Tar.

Tipping

As an alternative to dehorning when the threat of fly

strike is serious or where at least some of the horns are
required as for work oxen, the outer tips of the horn can
be removed. This will blunt the horn so less damage is
caused by fighting and will reduce the danger to the workers. Depending on the length and shape of the horn, a
variable portion of the tip of the horn has no blood or
nerve supply and can be cut off at any time without danger to the animal or inducing bleeding. Tipping can be
done with any of the tools that are used for dehorning.
If no bleeding is induced, no treatment is necessary.

As a final word on dehorning it should be pointed out
that it is a difficult, dirty, bloody procedure that is
hard on the workers and on the animals. If it is at all
possible dehorning should be avoided by regular disbudding
of all calves (see Section 5.3).

3.7 SIGNS OF HEAT

The time of heat or oestrus in cattle is when the females
are receptive to the male. This is also the time when
ovulation occurs. When a bull is with a herd of cows all
the time, there is no need for the herd owner to detect
heat since the bull will find any cows that are on heat.
If, however, artificial insemination is used or if the
cows are taken to the bull only when they are on heat
(hand mating) it will be necessary for the owner to be
able to detect this condition. The signs of heat are as
follows:

The vulva is swollen.

There is a mucus discharge from the vulva.

The cow on heat will stand while other cows mount her.

Milk production may fall.

A cow coming onto heat will often be restless, spend
time mooing at bulls and may try to mount other cows.

N.B. A mounting cow is not necessarily on heat. It is the cow which stands to be mounted by other cows that is on heat.

Individual cows may show any or all of the above signs for variable lengths of time. Approximately two-thirds of all heats occur between 6 p.m. and 6 a.m.; i.e. during the evening, night and early morning. When observing to see which cows are on heat, it is thus best to observe early in the morning and late in the evening. Since the symptoms of heat may be affected by the presence of the observer, it is also necessary that the period of observation be at least 30 minutes. Feeding should not occur shortly before or during the period of observation since provision of food may interfere with the oestrus behaviour.

Some devices are commercially available for heat detection. One is a capsule of coloured dye which is affixed to the cow's back. This is broken when she is mounted by another cow which leaves a dye mark on her back identifying her as being on heat. It is also possible to make bulls incapable of making fertile matings. This can be done by removing a section of the ductus deferens which carries the sperm from the testes (vasectomy) or by moving the location of the penis so the bull is incapable of mating the cow (penectomy). Both of these operations must be done by a veterinarian. Such bulls can safely run with the cow herd and, if they wear a marking harness on the chin or brisket, will place a mark on any cow which is on heat. Steers are not suitable for this since removal of the testes also removes sexual interest.

Whatever system of heat detection is used, regular and careful observation and recording is necessary to be sure that every cow is noted every time she is on heat. This allows regular and timely service and results in regular calvings for all the cows.

3.8   PREGNANCY DIAGNOSIS

Diagnosing pregnancy in domestic livestock can be

accomplished by several means. The easiest method is
visual observation of the cow for obvious external signs
of pregnancy such as an enlarged abdomen, enlargement of
the udder and swelling of the vulva. This method is normally most effective in the last month or two of pregnancy
but with experience it may be possible to determine pregnancy visually much earlier. Absence of oestrus may be a
sign of pregnancy but may be due to other causes so the
non-cycling cow should also be checked by another method
to confirm pregnancy.

If one exerts pressure low on the abdomen just behind
the ribs and pushes the hands upwards toward the back, it
is often possible to 'bump' a calf after the third or
fourth month of pregnancy when the pregnant uterus moves
down into the abdomen. This technique is knows as bal-
lottement.

There are several methods of determining pregnancy in
the laboratory using biological or chemical methods.
These methods are normally not practical for the livestockman as they are expensive and require special care. They
might be practical for an extremely valuable dam but not
for the regular, day-to-day livestock operation.

The most commonly used method of pregnancy diagnosis
in cattle is palpation of the uterus and reproductive
tract through the rectal wall. In skilled hands this is
a rapid and highly accurate method which allows the stage
of pregnancy to be determined as well as deciding whether
or not the cow is pregnant.

Technique for rectal palpation

1. Thoroughly wash the perianal region of the cow with
soap, water and antiseptic.

2. Don a palpation sleeve and lubricate well with soap.
Be sure the fingernails on the palpating hand are trimmed
short and smooth and that all rings, bracelets and watches
are removed. The lining of the rectum is very tender

tissue which can be easily damaged and result in severe
bleeding.

3. Form the fingers of the palpating hand into a cone and
gently force the fingertips into the rectum.

4. Spread the fingers to allow air to enter the rectum.
This will tend to cause relaxation of the anal sphincters.
If the cow humps her back, having your assistant scratch
the top of the back will also often promote relaxation.

5. Insert the hand about wrist deep into the rectum and
remove any faeces which may interfere with palpation.

6. With the hand wrist deep into the rectum feel downward
and locate the floor of the pelvic arch. Either on or
just anterior to the arch you should locate the cervix
which is a solid, muscular organ about 100 mm long and
about the size of a maize cob.

7. Palpate anteriorly along the cervix to the body and
then to the horns of the uterus.

8. Carefully check the horns of the uterus for the signs
of pregnancy as described in Table 11.

Pregnancy diagnosis by rectal palpation is a skill
acquired through practice. The most difficult aspects are
diagnosis prior to three months and actually determining
the stage of pregnancy. Both require palpation of numerous cows with comparison against actual calving or breeding records. On the other hand, to tell if a cow which
has been served at least three months previously is, or is
not, in calf is not difficult and is a useful skill for a
livestockman to have. You should, however, try to obtain
training under a skilled person before attempting diagnosis on your own as improper handling may result in injury
to the cow or to the foetus.

Table 11. *Signs of pregnancy by rectal palpation.*

Stage of pregnancy (months)	Signs
Not pregnant	Uterus and cervix on pelvic arch; both horns same size.
1	Slight enlargement of pregnant horn; may feel extra-uterine 'slip' due to foetal membranes.
2	Pregnant horn enlarged; definite 'slip'; may feel amnion (10 mm diameter at 5 weeks to 40 mm diameter at 7 weeks).
3	Uterus moves over anterior rim of pelvic arch; 'fremitus' may occur; definite 'slip'.
4	Uterus further into abdomen; can palpate foetus and foetal-maternal attachments; 'fremitus' may occur.
5	Uterus down into abdomen with cervix just at pelvic rim; may palpate head or legs of foetus or it may be out of reach.
6	Foetus may be out of reach or may be able to 'bounce' the head like a basketball; mammary glands of heifers begin enlarging.
7	Same as at 6 months except udders of all cows will begin to enlarge.
8	Foetus can again be palpated.
9	Foetus moves further up from floor of abdomen; udder enlarged and appears waxy; may observe vulval mucus discharge.
9 to parturition	See Section 3.9.

Adapted from Hafez, E.S.E. *Reproduction in Farm Animals*, 2nd ed. Lea and Febinger, Philadelphia, 1968

3.9   PARTURITION

Parturition means the act of giving birth. This section will discuss the events which occur prior to and during birth as well as some recommended procedures for the new-born calf and some problems which may occur.

Parturition

It is not known exactly what prompts the cow that it is time to give birth nor exactly what factors coordinate all the birth processes. As the calf grows in the abdomen and reaches the normal size for birth its orientation will normally change to the position necessary for birth as shown in Fig. 35. Changes in the cow occur before this, however, with enlargement and growth of the mammary gland in heifers beginning as early as the sixth month of pregnancy. Enlargement of the udder generally begins in the late seventh or early eighth months. In the latter part of the ninth month the teats and udder will become hard and appear waxy. Some swelling of the vulva also occurs.

In the day or two prior to calving hormones from the brain will cause the pelvic ligaments to soften so that the pelvic arch can enlarge to allow passage of the calf. This softening of the ligaments of the pelvis decreases the degree of connection between the two sides of the pelvis so the cow will appear to have a rolling gait and her hindquarters will sway from side to side as she walks. Depressions will also appear just anterior to the pin bones and the tail head will be raised. During this period there is often a mucus discharge from the vulva.

During the last few hours before calving the cow will appear restless and may wander away from people and the rest of the herd seeking a secluded place to calve. A cow that is calving should not be disturbed. A very large percentage of all parturitions in cattle occur without the need for any human interference. The presence of people at this time may upset the cow to the extent where she will not calve normally.

Dystocia

There are times when assistance is required. The word
dystocia is taken from the Greek and means difficult birth.
Several factors may contribute to difficulty in parturition including congenital abnormalities of the young, extremely small size of the mother and/or large size of the
calf or abnormal placement of the offspring in the uterus.

In the first two instances, it is often necessary to
dismember the foetus or to remove the foetus from the
uterus by Caesarean section. Both procedures require a
high degree of skill and should be performed by a veterinary surgeon whenever they are necessary.

Figure 35 depicts a calf in the uterus of the cow in
normal position for delivery, i.e. with both front legs
forward and the head lying on them. The hind legs are below the calf so that, as the calf moves through the birth
canal, they will be thrust out backwards. This is important as it flattens the hips of the calf so it may pass
more easily through the canal.

Abnormal positions in the uterus include calves which
are upside down or backward, have one or both hind legs
forward or one or both front legs back, or the head may be
turned back.

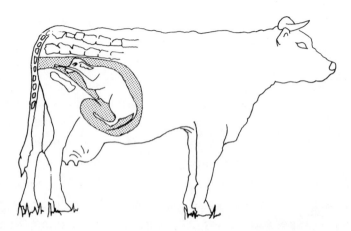

Figure 35. Normal position of the calf just prior to parturition.

<u>Treatment of dystocia.</u>   Unless the stockman has some ex-
perience or it is an emergency, it is best to call a
veterinarian if a cow has been attempting to calve for six
to eight hours and has been unsuccessful. If a veterinarian is unavailable, the following are guidelines which may
be of use.

Observe strict standards of cleanliness at all times
(see 1.9). Introduction of germs into the uterus may
result in severe metritis which can be very difficult
to cure and may result in infertility or the death of
the cow.

Confine and restrain the cow so that she cannot hurt
herself or the workers. If she is tied, make sure it
is a low tie or a long tie so she will not be suspended by her head if she should go down.

Check the amount of lubrication on the vaginal walls.
After the feet and head of a calf have been in the
vagina for some time, the vagina tends to dry out and
provides even more resistance to expulsion of the
foetus. If the walls are dry, lubricate them with
soapy water or a neutral oil such as medicinal paraffin. (N.B. This is not the paraffin or kerosene used
as a fuel for lamps.)

Gently insert your hand into the vagina to check the
position of the calf and its limbs. If there is a
limb out of position, gently push the foetus back into
the uterus before attempting relocation. This may be
impossible due to uterine contractions. If this is
the case it will be necessary to obtain veterinary
assistance to administer anaesthetic to stop the contractions.

If no malposition is found, you may be able to assist
the cow by placing tension on the calf by pulling on
its legs (by use of a rope or chain) or its head (by
means of blunt hooks placed in the eye sockets). The
chain is preferable as it is easier to sterilise than
rope but take care that the chain does not cause damage to the calf's legs.

You must not pull the calf. Your function is to maintain
tension after each 'push' by the cow.

The angle at which you exert tension is important.
Until the head and shoulders have been expelled, you

should pull nearly straight back from the cow with only a slight angle toward the cow's feet. After this, try to maintain nearly the same angle the calf would take if the cow were calving standing up. This is important because of the necessity of maintaining the proper angle of the hips to facilitate their passage through the pelvic arch.

If the head is in the vagina so the calf appears to be caught by the shoulders in the pelvis, pulling one leg, then the other may free the shoulders. If the problem is with the hips, rotating the calf by 90° may free the calf.

Do not forcibly pull a calf from an exhausted cow i.e. one which has ceased trying to push the calf out. Forcible pulling is very likely to cause fatal internal injuries to the cow. You may find however that, though the cow appears exhausted and contractions have ceased, as soon as she realises that you are assisting her, contractions will start again. If the cow does not respond to assistance it may be necessary to obtain veterinary assistance to remove the calf. Do not waste too much time trying to deliver a calf yourself. If you are unable to do so within 30 minutes or so, it is unlikely you will be successful and the longer you delay, the more difficult it will be for the veterinarian.

Care of the newly born calf

When a cow calves normally, she will almost immediately begin licking the calf. This serves to dry the calf and also to stimulate breathing and the muscles so the calf will get onto its feet and nurse. If the cow is exhausted she may not be able to stand to lick the calf so the calf should be placed by her head so she can lick it. If, when the calf is delivered, it is not breathing, remove the mucus and membranes from the mouth and nose. If there is liquid in the lungs, hold the calf upside down to drain it. Blowing gently into the mouth and nostrils may stimulate breathing or it may be necessary to slap the calf gently on the chest over the heart.

Some dams, especially first calf heifers, will refuse to acknowledge their calves and will not lick them or allow them to nurse. Sprinkling common salt or concentrates on the calf may get the cow to lick. If this fails it is advisable that the calf be dried off with wisps of hay or with sacks and the cow confined so the calf can nurse. This first nursing is important to the calf to provide energy and stimulus for moving as well as to acquire the antibodies which are contained in the colostrum. Check the udders of cows that are reluctant to allow the calf to nurse to be sure that there is no mastitis or other soreness. It may be necessary to milk out any excess milk after the calf has had as much as it requires. Cows seldom will disown their calves after they have suckled so once the calf has had a chance for the first meal, the cow and calf can be released. They should be watched, however, to be sure that the cow is feeding the calf.

If calving occurs in an area where other calvings have occurred or if assistance has been necessary during calving there is danger of infection to the calf through the navel. To overcome this it is useful to dip the navel in tincture of iodine. Simply place the iodine in a small cup and press the cup against the calf's abdomen ensuring that all parts of the umbilical cord are soaked with the solution.

Metritis

Metritis means an inflammation of the female reproductive tract. This condition may be either acute or chronic and may be throughout the tract or localised in one area e.g. cervicitis. In chronic cases, large quantities of pus may accumulate in the uterus. The causes and symptoms are diverse and the condition should normally be treated by a veterinarian.

Probably the most common cause of metritis in cattle is a retained placenta. The placenta is normally passed

within eight hours after calving. If it is not and remains in the uterus for several days the tissue may putre-
fy and lead to infection. Treatment for retained placenta
may involve placement of large pills known as pessaries in
the uterus and intramuscular injection of antibiotics.
The pessaries contain hormones which will stimulate expulsion of the placenta as well as releasing antibiotics to
combat infection. Under no circumstances should the placenta be forcibly pulled from the uterus. Gentle traction
may be applied but extreme caution must be used. There
are numerous blood vessels contained in the connections
between the uterus and the placenta and pulling the placenta may result in the rupture of these vessels and consequent haemorrhaging. As a general rule, removal of a
retained placenta should be left to a veterinarian who has
been trained to do this without harming the cow.

Metritis must be considered a contagious disease and
strict measures taken to prevent spread to other stock.
All discharges and material from the infected uterus
should be buried or burned. Scrupulous cleanliness of all
persons handling or treating the infected cow must also be
observed and all equipment must be cleaned and disinfected
after each use.

Metritis can be extremely difficult and expensive to
cure unless the proper treatment is initiated very early
in the course of the disease. It is strongly recommended
that a veterinarian be consulted as soon as you feel that
metritis might develop.

3.10   VACCINATION

Vaccination is the injection of material into an animal to
promote immunity to a disease. The efficiency of vaccines
is variable; some give full immunity for a life-time
whereas others may only promote an ability to tolerate a
disease for a short period. Whether or not to vaccinate
your stock against a specific disease depends on the
probability of your animals contracting the disease, the

efficiency of the vaccines available and the cost. Cost is seldom a real consideration if the first two indicate vaccination. Most vaccines are very inexpensive relative to the value of cattle and the cost of vaccinating a large number of animals is seldom as much as the value of one animal.

You should contact the Veterinary Department for advice as to which diseases you should vaccinate against and how to obtain the vaccines. In many cases, they will also do the vaccinations for you.

Table 12 indicates some of the diseases against which vaccination is commonly carried out, the frequency with which the vaccinations should be done, and the ages of the animals to be vaccinated.

Table 12. Common vaccinations for cattle.

(a) Vaccinations done once.

Disease	Age to vaccinate
Brucellosis (Contagious Abortion)	3-6 months (females only)
Salmonellosis	1-2 weeks
Heartwater	Birth

(b) Vaccinations requiring repetition.

Disease	Age to vaccinate	Frequency
Anthrax	All	Annually
Black-quarter	All cattle 6-36 months old.	Annually
Botulism	All	Annually
Foot and Mouth Disease	All	6 months as required
Lumpy Skin	All	As required
Tetanus	All	Annually
Rabies	All	Annually as required

Testing for tuberculosis and brucellosis

In areas where TB and brucellosis (CA) are endemic and in
dairy herds producing milk, all cattle should be tested
annually for these two diseases. This is a job for a
veterinarian but the following briefly describes how the
tests work.

Since the bacilli which cause tuberculosis in cattle
also infect humans and transmission is widespread by way
of exhaled air, contact, or milk, it is important that infected animals be identified. The tests in use for TB depend on the fact that infected animals are allergic to the
proteins contained in the solution in which the bacilli
have been grown (tuberculin) and react when exposed to
them. The tuberculin is injected into the skin (intra-
dermally) in one of the folds under the tail or in the
neck. Infected animals will show a characteristic lump
48-72 hours after injection. This may be caused by TB,
but brucellosis, Johne's disease or other non-specific
bacilli may also cause a reaction. To test this, the animal is injected in two sites: one with avian tuberculin,
the other with bovine tuberculin. The non-tuberculosis
bacilli will not react or react only slightly with both
the bovine and avian tuberculin whereas the bovine tuberculosis bacilli will give a noticeably stronger response
to the bovine tuberculin. Diagnosis thus depends on the
difference in the size of the swelling at the two sites.
This is known as the comparative intradermal test.

Brucellosis in cattle is caused by the bacterium
Brucella abortus. The disease in humans is known as
'undulant fever' and can be transmitted by contact or in
the milk. Treatment in humans is relatively ineffective
so it is important that all cattle producing milk for
human consumption be free of the disease. Two common
tests are the ring test and the serum agglutination test.

The ring test is based on the agglutination of milk
from infected animals and is usually utilised to check
entire herds. The serum agglutination test is used to

locate infected individuals within a herd. With the serum agglutination test there is no sharp dividing line between positive and negative so a third classification ('suspect') is used. Cattle vaccinated after 6 months of age with the S19 vaccine may fall into the suspect class throughout their lives. Normally, suspect animals are retested after a month. In some countries, if this second test also results in a suspect classification, the animal must be destroyed. Since this may be caused by late vaccination, it is important to vaccinate early. Because of the possibility of an animal giving a suspect reaction throughout its life as a result of the vaccination, bulls are not normally vaccinated. Since only a small proportion of the bulls born should be used for breeding, it is much better to test them frequently than to run the risk of having to destroy them as a result of the vaccination.

Some of the newer vaccines such as S45/20 do not interfere with the serum agglutination test so can be used at any age. Initiation of immunity with S45/20 requires two vaccinations six months apart.

4 Dairy Cattle

4.1 DAIRY TYPE

Dairy cattle are kept to produce milk so the important factors relating to dairy type are those which will promote high daily production through full lactation periods of about 10 months, a calf each year and the type of milk desired on the feeds available. The dairy cow with the above characteristics will be the most economical but all the desired characteristics are seldom found in one animal. It is difficult to tell by looking at a cow whether she has most of these properties but there are certain bodily signs which do indicate a dairy type animal.

The beef animal, particularly of the British breeds, will have a square blocky shape. In contrast, the dairy animal should have a wedge shape. This will be seen when looking from the side in that there is less depth of body at the front of the animal than at the rear. When looking from behind the animal down its back there will also be a wedge shape from the hip bones to the shoulder. The third wedge shape is seen when facing the animal as she should be wider at the points of the shoulders than at the withers. In addition to this triple wedge shape the dairy cow should have a graceful head and neck without coarseness or excess tissue around the brisket and dewlap. The barrel should be large and the ribs well sprung with a strong, straight topline, indicating a good capacity for food and water.

The bones of the dairy animal should be strong but not coarse. The legs should be evenly placed to provide the ability to travel for grazing as well as less tendency for lameness.

The udder should be large, well-balanced and have tight attachments to the body. The teats should be squarely placed on the four quarters and be of sufficient size for efficient milking.

Most breed associations publish exact descriptions of what characteristics the ideal bull and cow of that breed should have. Although it is useful to have this ideal in mind, it must also be remembered that many long-lasting, high-producing cows have what might be considered serious faults by the associations.

The diagram in Fig. 36 shows the general shape desirable for a dairy cow. Also indicated on the diagram are the names of the various parts of the dairy cow. As a stockman you should know and be able to correctly use these names.

Some common dairy breeds include the following:

| Ayrshire | German Brown | Holstein-Friesian |
| Brown Swiss | Guernsey | Jersey |

4.2 CLEAN MILK PRODUCTION

Milk is nature's most nearly perfect food, being deficient only in iron. As such, it is an excellent food for raising the young of any species but, at the same time, is an ideal medium for bacterial growth. At the point of production, i.e. the dairy, the milk is in close proximity to prolific sources of bacterial infection which can change milk from an ideal food to a potent disease vector.

The production of clean milk can be viewed from three aspects: (a) the disease status of the cow, (b) the cleanliness of the milking process and (c) the rapid cooling of milk to prevent or inhibit bacterial growth.

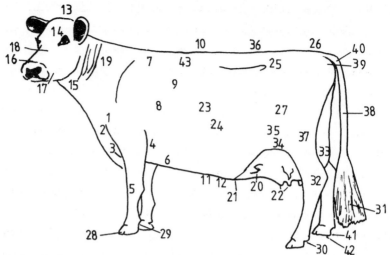

1. Point of shoulder; 2. Dewlap; 3. Brisket; 4. Point of elbow;
5. Knee; 6. Chest floor; 7. Withers; 8. Heart girth; 9. Crop; 10. Back;
11. Milk wells; 12. Mammary veins; 13. Poll; 14. Forehead; 15. Throat;
16. Muzzle; 17. Jaw; 18. Bridge of nose; 19. Neck; 20. Fore udder;
21. Fore udder attachment; 22. Teats; 23. Barrel; 24. Ribs; 25. Hip or
hook; 26. Rump; 27. Thurl; 28. Hoof; 29. Pastern; 30. Dew claw; 31.
Switch; 32. Hock; 33. Rear udder; 34. Flank; 35. Stifle; 36. Loin;
37. Thigh; 38. Tail; 39. Pin bones; 40. Tail head; 41. Heel; 42. Sole;
43. Chine.

Figure 36. Parts of a dairy cow.

All cows producing milk must be routinely examined for the
presence of disease. Most important are tuberculosis and
brucellosis (see 3.10) since both diseases can infect
humans and can be transmitted via the milk. It is important to avoid contaminating human milk supplies with milk
from cows with mastitis. This milk should not be fed to
calves or pigs either as it may cause scouring. Many
drugs, when administered to cows, appear in the milk within a few hours or days. Care must be taken that these
drugs do not contaminate human milk supplies. This
applies particularly to antibiotics to which many people
are allergic. Antibiotics can also spoil the milk for
cheese production.

Cleanliness of the milking process includes cleanliness of the cows, the humans involved, the buildings and
the utensils as well as a conscientious effort to keep
faeces, dirt, etc. out of the milk. All persons working

with the milk must be disease-free (including tuberculosis), must wash thoroughly prior to milking and should wear clean, white overalls. No one should be allowed to smoke within the confines of the dairy while milking is in progress or when uncovered supplies of milk are present. Buildings for milking and for handling milk should be roofed and screened to eliminate contamination by flying insects, dust and rain. All floors should be concrete and walls plastered or tiled so that the facilities may be washed with soap and water after every milking. During the milking process all utensils must be periodically rinsed with a sterilising agent and the utensils thoroughly washed after each milking (see 4.5). Before milking, the udder and teats should be thoroughly washed with a disinfectant solution and any dirt or faeces which might contaminate the milk removed. Each quarter of every cow should be regularly checked for mastitis using one of the detection methods described in Section 4.6. The milk from infected cows should be discarded. During the milking process, care must be taken that foreign matter does not come into contact with the milk.

No matter how thorough one is regarding the cleanliness of the milking process, bacteria will still be present in the milk and will rapidly multiply in the nutritious environment provided by milk. This growth can be retarded by cooling, so it is essential that the milk be cooled as soon as possible after it is removed from the cow. The cooling will arrest bacterial growth until the milk can be heat-treated or pasteurised to kill the bacteria.

The production of clean milk involves scrupulous cleanliness at all stages of handling. The point at which the milk is removed from the cow is probably the point of greatest potential contamination and is the point at which the greatest care must be taken. Most countries have regulations governing milk production and dairymen should familiarise themselves with them.

4.3 HAND MILKING

When budgeting for profits in dairying there is a rather
narrow margin of return. Efficient milking techniques are
as important as any of the other considerations to be
taken into account. A farmer must appreciate the fact
that the well-being of his enterprise depends on the proper and efficient milking of his cows day after day without a miss. In South Africa, for instance, the average
number of cows hand-milked per man-hour is 5, but European
figures give an average of 8 per man-hour. This is largely due to the method of preparation and milking.

Preparation for milking

Whether you are hand or machine milking, the cow must be
adequately prepared in order to remove all the milk from
the udder. Milk formation is a continuous process in the
udder. As the milk is formed it is stored in the udder
and is unaccessible for the calf or milker unless certain
physiological events occur.

This is called the milk let-down and is mediated by a
hormone from the brain which is released when the cow receives a stimulus indicating that milking is about to
occur. The original stimulus for this was the presence of
the calf and its nudging of the udder. With domestica-
tion, cows have been trained to let down their milk when
the udder is washed or when it is brought into the milking
area. Some cows however may require the presence of the
calf for let-down. Let-down may be inhibited if the cow
is frightened or upset before or during the milking process.

Let-down is important as it allows the cow to be
thoroughly milked. This is important from the standpoint
of profit as well as the prevention of mastitis. Once the
hormone has been released, its effect will last for 5 to 8
minutes. It is important that milking be completed within this time so all the milk can be obtained.

Strip milking (Fig. 37(b)) is commonly practised in Southern Africa where the teat is held between thumb and first (index) finger. The milk in the teat is rubbed out by pulling the hand downward along the teat. Pinch milk-ing (Fig. 37(c)) is done by pinching the teat between the first and middle fingers, and stripping downwards. Both of these methods tend to be injurious to the tender tissues and may result in development of scar tissue in the teat cistern. The pulling of the udder will also re-duce its effective life by tearing the tissues of the udder and making them more prone to bacterial invasion. Milking salve must be used with either of these methods.

A more efficient method is illustrated in Fig. 37(a), and involves the rhythmic squeezing of the milk from the teats with the minimum of pulling on the udder. This is done by holding the teat with the thumb and first finger encircling the upper part of the teat and the remainder of the fingers in close contact. The hand and teat move up and down together but no rubbing action takes place. The milk is squeezed out by tightening the fingers starting with the index finger. The second, ring and little fin-gers are then sequentially tightened. All fingers are

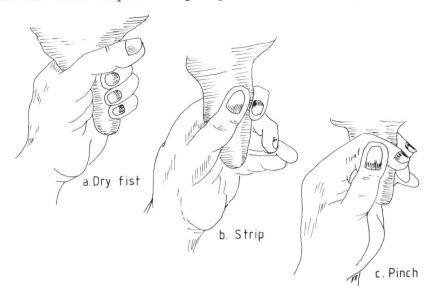

a. Dry fist

b. Strip

c. Pinch

Figure 37. Methods of hand milking.

then relaxed simultaneously and a new sequence begun. Dry
fist milking at first seems slow and inefficient but with
practice can be much faster by far than either of the
other two methods. It also produces cleaner milk since no
lubrication is required and the milk does not come into
contact with the milker's hands.

4.4 MACHINE MILKING

Whereas with hand milking the milk is squeezed out of the
teat, the milking machine is designed to suck the milk
out. To do this, a vacuum is maintained within the milk
receptacles and its connections to the teat liner which
surrounds the teat. Continuous vacuum on the teat would
shut off the blood supply, however, and cause discomfort.
To avoid this, the area outside the teat liner is alter-
nately at atmospheric pressure and under a vacuum. Since
there is always a vacuum inside the liner, atmospheric
pressure outside the liner will cause the liner to collapse and close below the teat. This cuts off the vacuum
to the teat and allows the blood to flow in the teat.
This is known as the massage phase. When a vacuum exists
outside the liner, the liner expands and the vacuum within the liner sucks the milk from the teat. This is known
as the milking phase of the cycle.

The change-over between the massage and milking
phases is controlled by the pulsator. The rate of change-
over (the pulsation rate) will normally be from 40 to 60
times per minute. The milking ratio is the duration of
the milking phase in relation to the massage phase and
can vary from 1:1 to 3:1 depending on the make of milking-
machine used.

The vacuum used for milking is caused by the vacuum
pump which evacuates air from the system. For the system
to function properly, it is important that the vacuum be
constant and at the right level. This is achieved by the
vacuum controller which allows air into the system to
maintain the correct, constant vacuum. This is approxi-

mately the vacuum created by a calf sucking and is equivalent to 0.5 bar (about one-half atmospheric pressure).

The milk claw is usually equipped with a shut-off valve which shuts off the vacuum to allow painless removal of the claw and to keep the unit from sucking dirt should it fall on the floor.

Steps in machine milking

Preparation and foremilking.
Wash and massage the udder. This causes let-down of milk so should be done thoroughly and gently. Milk 2 or 3 squirts of milk into the strip cup to check for mastitis (see 4.6).

Attachment.
Attach the teat cups about one minute after the beginning of preparation. This allows time for let-down of the milk. Attach the machine smoothly and gently, avoiding the intake of unnecessary air into the system.

Machine milking.
Most cows will milk out within five minutes. Hard milkers which always require a longer time should be culled from the herd.

Final check.
You should carefully watch for the cessation of milk flow and keep the machine on the cow for the shortest possible time. If the machine is left on too long it may cause irritation to the udder which could lead to mastitis. To be sure you have not left some milk, you should machine strip the cow. This is accomplished by gently pressing downward and forward on the claw to release any milk which may have been pinched off by the presence of the claw. Hand stripping is unnecessary and undesirable as it tends to turn the cows into slow milkers.

Removal.
Shut off the vacuum at the line and release the vacuum at the shut-off valve on the claw. Let a little air in between one teat and the teat liner and gently remove the unit from the cow. The machine should

never be pulled from the udder without breaking the vacuum.

Machine milking is an efficient method of milking cows which will result in maximum production and a minimum of health problems if used properly. If used improperly, however, machine milking can severely reduce milk yields and result in serious health problems. Therefore, the strength of the vacuum and the pulsation ratio should be checked regularly, and care taken that the proper procedure is followed at every milking.

4.5   WASHING MILK EQUIPMENT

Thorough cleaning of all milking equipment after every milking is absolutely essential for the production of clean, wholesome milk. There are four types of soiling which must be removed from the equipment: fresh, wet milk residues, air-dried milk residues, foreign matter, and heat-dried residues.

Fresh, wet milk residues should be removed from the equipment as soon as milking is completed by rinsing the equipment in warm water. Warm water is best as hot water may cause denaturation of the milk proteins and cold water may cause the fat residues to congeal. If cold water is used for rinsing, much larger quantities must be used.

Air-dried milk residues are due to spillage on the outside of equipment or result from improper rinsing after milking is completed.

Foreign matter includes dust and other particulate matter, faeces, dirt, etc. This matter is not normally soluble and most should be removed by the initial rinsing.

Heat-dried residues do not occur to any extent on the farm since they are primarily due to processes involving heat treatment of milk, e.g. pasteurisation.

The objective of any washing programme should be to remove all the above types of soiling, to sterilise the equipment and to promote the long life and usefulness of

the equipment. The following is a basic handwashing programme which is applicable for hand milking equipment or
bucket type milking machines.

1. Immediately after milking, flush the system thoroughly
with warm water. Remove external dirt by scrubbing and
rinsing with warm water.

2. Completely dismantle all equipment and scrub all parts
in a hot (46°C) detergent-steriliser solution. Follow the
manufacturer's instructions regarding the use and concentration of the solution. Be especially careful that all
rubber parts are thoroughly cleaned. Total contact time
with the solution should not be less than two minutes.

3. Rinse all parts with clean water or a halogenated
rinse solution.

4. Hang the parts on a rack where they will drain and dry
thoroughly. (Alternatively, the rubber parts may be
stored in a suitable alkaline detergent-steriliser solution or a halogenated steriliser.)

Rubber parts of milking machines are rapidly destroyed by
alternate wetting and drying. In addition, they absorb
butterfat which causes further deterioration. Regular
treatment by soaking in a 10% caustic soda (lye or sodium
hydroxide) solution will prolong the life of these rubber
parts. Manufacturers of milking machines will provide
instructions regarding care of rubber parts. These should
be carefully followed.

Due to faulty washing of the milk equipment or very
hard water, a scale known as milkstone will periodically
build up on milking equipment. This is best removed by a
commercial acid milkstone-remover. The equipment should
be washed in the usual way after milking and then soaked
for 5 to 6 minutes in the acid solution. After soaking,
the milkstone can be scrubbed and rinsed off. The equipment should then be rewashed to remove all traces of the

acid. Removal of milkstone need not be done daily but should be carried out as necessary.

Fouled vacuum lines are a more common source of contamination than is generally realised, so the lines used for milking machines should be cleaned at monthly intervals by drawing a detergent-steriliser solution into them. They should then be flushed with hot water to remove the detergent.

The maintenance of clean milking equipment depends on the use of a proper system of cleaning which is carefully carried out at each milking. If such a system is established and regularly used, an important potential source of contamination of milk can be eliminated.

4.6 MASTITIS

Mastitis is a general term which covers a variety of bacterial and fungal infections of the udder. Since the causal agents are ever-present, all dairymen will at one time or another need to deal with the disease. Many factors predispose to mastitis by allowing the bacteria and/ or fungi to enter the udder. Among these are bruising, chilling or over-filling of the udder, improper milking, etc.

Mastitis may be chronic or acute. The chronic form is the most common and may persist for months with no external signs of the disease and then may suddenly flare up. Then one will see lumps or clots in the first few drops of milk taken from the infected quarter. If untreated, the size and number of the clots may increase and the udder may be swollen and painful causing the cow to be lame in the hind leg of the infected side. As the disease progresses, the quantity of milk produced will decline and the milk will become thin and watery. Eventually permanent damage may result in a 'blind' or non-milking quarter.

Acute mastitis may be due to a flare-up of chronic mastitis or may be due to a recent infection. The infected quarter will be swollen and painful; the cow will have an elevated temperature and may refuse food, may scour, or lose condition and eventually die.

Detection of mastitis

The commonest and easiest method of mastitis detection is through the regular use of a strip cup. This is a cup with a fine screen or bakelite plate onto which the fore-milk (the first two or three squirts from each teat) is expelled. If the cow is infected, examination of the screen or plate will reveal the presence of the clots or floccules mentioned previously. A piece of dark cloth over a teacup can also be used as a strip cup.

Since mastitis milk is usually more alkaline than nor-mal milk, a colour change which signifies altered acidity can be used for testing for mastitis. Special cards for this purpose are available. Each card has a spot of colour for the milk from each teat. The foremilk from each teat is dripped onto the respective spot and each spot examined for the colour change characteristic of mas-titis. This test is especially useful diagnosing chronic cases of mastitis.

The California Mastitis Test is based on the fact that mastitis milk normally has an elevated cell count which will lead to a precipitate in the presence of specific chemicals. Equal amounts of the milk and reagents are mixed. If mastitis is present a precipitate will form. This test is not normally used in a day-to-day routine as it is slower than the use of a strip cup or blotter. It is useful for weekly or monthly checks, however, and the results tend to correlate with the result of microscopic examination.

Finally, the presence of mastitis can be determined by microscopic examination of the milk samples or by bac-terial culture of the milk. These tests are useful in the diagnosis of the causative agents in those cases which do not respond to normal treatment and may require special treatment.

Prevention of mastitis

The following procedures are useful for preventing the incidence and spread of mastitis.

The udders of all cows should be washed in a disinfectant solution before each milking.

All equipment used for milking, e.g. washing cloths, strip cups, buckets, should be thoroughly washed and disinfected between milkings. Strip cups should be disinfected whenever a cow with mastitis has been detected. Ideally the strip cup should be disinfected after every use.

The hands of the milkers should be washed and disinfected between cows. This is especially important when hand milking.

Treat all wounds on the udder promptly.

Treat cases of mastitis promptly and milk infected cows last.

Checking for mastitis before every milking is imperative.

Use a long-acting antibiotic at drying off to help prevent mastitis in the next lactation.

Treatment of mastitis

Milk infected cows last.

Use an antibiotic or combination of antibiotics (penicillin, streptomycin and/or aureomycin) infused into the infected quarter. In order to use the proper antibiotic and to avoid the development of resistant strains, it is advisable to send a sample of milk to a veterinary laboratory so they can determine which is the most effective antibiotic in that particular case. Administration of the antibiotic should proceed as follows:

Milk out the infected quarter

Thoroughly clean and disinfect the hands, the teat and the applicator

Insert the applicator fully into the teat

Expel the antibiotic into the teat

Work the antibiotic up into and throughout the quarter by massaging.

Chronic cases are usually treated two or three times at intervals of 48 to 72 hours. Acute cases should be treated at intervals of 12 to 24 hours until the symptoms subside. In no case should milk from infected quarters be used for human or livestock feed within 72 hours after the last treatment because of the antibiotic in the milk.

In very severe cases, injection of penicillin or another antibiotic intramuscularly may provide some benefit but this should be decided by a veterinarian, who should be called if the cow does not quickly respond to treatment with intramammary suspensions. Do not delay: mastitis may be fatal.

5 Calves

5.1 CALF MANAGEMENT

Calves represent expansion of your herd as well as a substantial source of income. Calves are quite hardy and, if
given proper food, housing and care, are generally able to
perform quite well even when fancy facilities are not
available. The simplest method of calf rearing is to
leave the calf with its dam. During this period the calf
will receive most of its food from the cow in the form of
milk. This is the common practice in herds where the
cattle are kept only for beef production. If, however,
milk is desired for home use or for sale it may be necessary to remove the calf from the cow within a few days of
birth and hand-rear it.

Whichever method of rearing is used, it is important
that the calf receives the colostrum from the cow just
after birth. This milk is higher in energy and vitamins
than normal milk and contains antibodies which are essential for the calf's survival. Generally the newborn calf
will nurse the cow within three to six hours. After this
first feeding, calves to be hand-reared can be removed
from the cow. For the next three days, however, the milk
from the cow should be fed to the calf by means of a
bucket. Other practices which should be carried out with
all newborn calves include disinfection of the navel (see
3.9) and identification (see 2.3 to 2.9).

Housing for calves need not be extensive. The calf needs protection from direct winds, direct sun and getting wet but is able to withstand quite extreme temperatures. Housing should be constructed so that it is easy to work with the calves and easy to clean when no calves are present. Calves up to 150 kg require 1.3 to 2.0 m^2 pen space. This is especially important as some of the diseases of calves can be passed from one calf to another through the use of dirty facilities or equipment.

Although young calves will eat very little grass, the opportunity to exercise in a clean pasture during days with good weather is a beneficial procedure with hand-reared calves. The pastures used for calves must not be used continually and should not be used for adult stock because of the build-up of internal parasites. Shelter from the sun should be available in the form of trees or artificial sunshades. Having the calves out of the pens allows for regular, thorough cleaning.

Calves are sometimes born with extra teats. These generally are not functional and do no harm but are usually removed from heifers that are going to be used as dairy animals as these teats distract from the neat, balanced appearance of the udder and, in some cases, may be in the way during milking. Removal is carried out by cleaning and disinfecting the area around the extra teats and cutting them off with a pair of sterile scissors. If this is done before the calf is a week old there will be little bleeding, if any. The wound may be treated with Stockholm tar or an aerosol wound spray.

The faeces of newborn calves are very sticky and may build up under the tail and block the anus so that the calf is unable to defaecate. If this happens, the faecal material should be softened with warm water and removed.

5.2   CALF CASTRATION

Castration of bull calves not intended for breeding purposes is highly recommended. This practice allows the

best bulls only to be kept for breeding and makes the
other animals easier and safer to handle. Bulls grow faster than steers but produce a carcass with less fat and
darker meat. Bull carcasses are often downgraded at the
market unless they are quite young. It is best to castrate calves as soon as the testes descend into the scrotum as the small calves are easier to handle and they
suffer less from the operation than larger animals.

Castration using a knife

When using a knife for castrating, an incision is made in
the scrotum and the testes are completely removed. Before
castrating, the operator's hands, the instruments and
scrotum should be cleaned and sterilised. Instruments
should be boiled and tissues thoroughly washed with soap
and water and rinsed with tincture of iodine or alcohol
(see 1.9). The actual castration procedure is as follows.

The calf should be flanked and securely tied.

Make a slit in the side of the scrotum over one testis.
An inner, shiny layer over the testis will then be
visible. Gently cut this and slide the testis out
through the incision.

The testis is supplied with a large artery so it is
important that the cord be severed in such a way as to
stop bleeding. This can be done by scraping with a
knife, using an emasculator, or by the use of a hot
iron.

It is possible to remove both testes through a single
incision, but it is preferable to make a vertical incision over each testis and remove them individually.
When the incision has been made, the testis should be
pulled from the scrotum and turned several times to
twist the cord. The knife should then be scraped
along the cord so the artery will be severed with a
ragged end. This helps to stop the bleeding.

An emasculator is an instrument with two sections;
one for cutting, the other for crushing. The crushing
edge should be toward the body so the cut occurs below
the crushed ends. Generally, emasculators are constructed with a wing nut on the cutting side. If so,

this should be toward the testis when the instrument
is used.

When using a hot iron to stop bleeding, the artery
should be pinched off with forceps near the body before the cord is cut. After cutting, the iron should
be applied to the cut ends before the forceps are
released.

Once removal of the testes has been completed, treat
the incisions as wounds i.e. be sure they are clean,
well drained, and that antibiotics are provided (see
2.19). For drainage purposes, rapid closure of the
incisions is undesirable. It is better to put the
calf on clean pasture rather than to keep it penned.
This will allow the calf to exercise, which promotes
drainage and reduces the chance of infection. As a
general rule, pastures tend to be cleaner from the infection standpoint than pens. This is because pens
are often used to house sick animals.

Castration using a Burdizzo

The Burdizzo is essentially a large clamp that is used to
crush the blood and nerve supply of the testes. There is
no open wound but the testes will degenerate due to the
lack of blood and nerves. To use a Burdizzo:

Flank and securely tie the calf.

Pull the testis into the scrotum and work the cord to
one side.

Position the Burdizzo over the cord and clamp it shut.
Hold for 30 seconds.

Remove the Burdizzo and do the other side.

Several points must be borne in mind when using the
Burdizzo:

Never crush both cords at once and be sure the two
crush marks are not aligned across the scrotum. In
either of these cases, the blood supply to the scrotal
skin may be damaged and the scrotum may fall off.

Be sure that the sigmoid flexure of the penis does
not get caught between the jaws.

With large animals, it may be necessary to clamp twice on each cord to be sure that castration is complete.

The Burdizzo is not 100% efficient and some animals may escape castration. It is therefore advisable to palpate the testes of all animals two to three months after castration. By then, animals which are properly castrated will have small, soft, or oddly shaped testicles when compared to uncastrated animals of the same age. Any animals which escaped castration the first time should be re-done (and re-checked in two or three months).

If you find that a large proportion of the animals were not properly castrated, check your Burdizzo to be sure it is adjusted properly. A properly adjusted Burdizzo should cut a piece of twine folded in paper without cutting the paper. One common cause of poor adjustment is storage of the instrument in the closed position. This is ill-advised as it may lead to fatigue of the jaws of the instrument so they will not close tightly enough to crush the cord.

Castration using elastrator bands

The elastrator is an instrument used for stretching a strong rubber ring which is then placed around the scrotum above the testes. The ring cuts off the blood supply to the testes and scrotum and results in their eventually sloughing off.

The disadvantage of this method is that it causes pain to the animal over a long period of time. It also may result in a wound which will be extremely slow healing and is exposed to infection. It is, however, very simple and sure, requiring no sterilisation. Castrates should be watched carefully for the development of infection.

Many countries have regulations governing the ages at which particular methods of castration may be used on animals. In Great Britain, for example, the elastrator is prohibited on animals over one week old.

5.3   DISBUDDING

The reasons for removing the horns of livestock have been pointed out in Section 2.6. It is easier to prevent horn growth in calves than to dehorn larger animals and the animals will suffer less of a setback. The process of preventing horn growth in calves is known as disbudding i.e. the removal or killing of the horn bud. This can be done by chemical methods, use of a hot iron, or surgical removal of the bud. All three methods destroy the area of specialised epithelium from which the horn develops.

Disbudding with caustic paste

There are various caustic pastes on the market which can be used to kill the horn bud. These are simple to use in that they are merely smeared on the bud area when the calf is 1 or 2 days old. However, if the pastes are carelessly applied or if the calf gets wet and the paste runs so it comes in contact with the eyes, permanent eye damage or blindness may result. Also, the caustic pastes are frequently not entirely effective and rudimentary horns or 'scurs' may result. Using caustic pastes is a method which is not recommended.

Disbudding with a hot iron

Although it requires more equipment and preparation than the use of caustic paste, a hot iron is more effective. The iron used has a depression which covers the horn bud so that a complete ring may be burned in the skin around the base of the horn. The calf should be flanked and restrained. The area over the bud is clipped free of hair and covered with a layer of petroleum jelly. The hot iron is then placed firmly over the bud and a well-defined ring burned into the skin around the bud. This serves to kill the nerve supply to the horn and prevent growth. If a commercially produced iron is not available, a 1 m length of electrical conduit makes an excellent substitute. It

should be bent into an L-shape and the handle portion
wrapped to prevent burning the operator's hands. Since
the thermal mass of such material is low, the iron should
be used immediately after removal from the fire as it will
cool very quickly.

Irons may be heated in a fire, with gas cylinders or
by electricity. When an electrically heated iron is used,
care must be taken to ensure that it is properly earthed.
Calves are much more readily electrocuted than humans.

Surgical removal of horn buds

For surgical removal of the horn bud, one should use a
scoop or tube disbudder specifically designed for this
purpose. After clipping the area around the horn bud and
disinfecting it, the scoop is used to cut out the entire
bud. If available, an injection of a local anaesthetic
into the nerve as described in Section 3.6 will lessen the
trauma suffered by the animal. This method is easy and
quick but does leave an open wound which must be treated
as such (see 2.19).

Some stockmen prefer to use a hot iron and then to
gouge out the bud within the burned ring. This provides
complete removal of the horn bud. Complete removal is
especially desirable for high quality purebred stock.
Here again, local regulations against cruelty to animals
must be observed. For example, it is forbidden by a number of countries to operate in this way on an animal over
one week old without using a local anaesthetic.

5.4 CALFHOOD DISEASE

The expansion and improvement of your herd depends on
growing as many calves as possible as rapidly as possible.
This means proper feeding, low mortality and as few checks
or set-backs as possible. Mortality or set-backs may be
a result of disease. A few of the more common diseases
are mentioned in this section but it should not be
assumed that these are by any means the only, or even the

most important, calf diseases as the importance will depend on the particular situation. In all cases, veterinary assistance is recommended; especially for diagnosing specific conditions.

Symptoms of disease

Symptoms of disease in calves may include:

> Elevated temperature
> Mucus discharge from the nostrils
> Coughing or sneezing
> Laboured breathing
> Scours (diarrhoea)
> Dull appearance
> Staring coat
> Refusal of food
> Listless or aggressive behaviour
> Lameness
> Shivering
> Jerky movements
> Over-reaction to handling
> Excess salivation

Whenever a calf is seen with one or more of the above symptoms or behaves or appears in any way abnormal, it should be isolated from other livestock in a clean, warm, dry pen. Veterinary assistance should then be obtained to diagnose the condition and prescribe treatment.

Calf diseases

Scours. Scours in calves may be caused by wrong feeding, worm infestation or infectious disease.

Wrong feeding may include too much milk, cold milk, irregular feeding or dirty buckets. The faeces will usually be white or a very light tan. The calf will seldom have an elevated temperature. Treatment involves the removal of the cause followed by severely reduced feeding levels with a gradual return to normal levels over a period of several days.

Worm infestation, especially intestinal roundworms such as *Haemonchus* spp., will result in watery, black scours. This colour is due to the digestion of blood

which is released due to the worm infestation in the upper
intestinal tract. Treatment involves ridding the calf of
the worms which are causing the irritation. If the calf
is weak, an injection of vitamin B12 (cobalamin) may be of
value.

Diseases for which scours may be symptomatic include
coccidiosis, white scours and paratyphoid. With coccidiosis, the faeces will be bloody and slimy whereas in the
case of white scours the faeces will be putty-like and
yellowish-brown. Gas bubbles and specks of blood may be
present. This is distinguishable from nutritional scours
because a calf with white scours will be feverish. Calves
with paratyphoid have a temperature of 40.5°-41.5°C. A
day or so after the onset of the disease, scouring develops which is characterised by typical foul-smelling yellow
faeces. The faeces may become bloodstained after several
days.

In all cases of scours, there is water loss from the
body which should be replaced by offering the calf small
quantities of cooled boiled water at frequent intervals.

Calf pneumonia.   This condition (also known as enzootic
pneumonia or virus pneumonia) is caused by a wide range of
viruses and often is complicated by secondary bacterial
infections. It normally affects only calves up to about
4 months of age although older animals may act as carriers.
Predisposing factors include contact with adult animals or
infected calves, overcrowding, dirty pens, exposure to
inclement weather and poor feeding conditions. Symptoms
are mucus discharge from the nose, elevated temperature,
laboured breathing and a dull or weak appearance. Treatment with properly prescribed antibiotics is usually
effective. In advanced cases, injectable vitamins may be
used.

Ringworm.   Ringworm (dermatomycosis) is caused by the
fungi *Trichophyton* and *Microsporum* and affects cattle
(especially calves), dogs, cats and humans and occasion-

ally sheep and swine. The fungus is transmitted by spores which lodge in skin cracks when animals rub against one another or come in contact with infected fences, bedding, etc. Ringworm is more commonly seen in rainy seasons than dry. In calves, lesions are normally found on the head and neck or rump regions. The fungus causes breakage of the hair at the skin surface and the skin takes on a grey, scaly appearance in round, sharply defined areas. The lesions itch so the animals spend much time rubbing. Definite diagnosis can be easily made by microscopic examination of scrapings of the skin from the affected areas. Such scrapings should be sent to a diagnostic laboratory as soon as the symptoms are noted.

Treatments for ringworm include tincture of iodine, 5% salicylic acid ointment, 2% formalin and various mercury-containing ointments. Before application, the areas should be scrubbed with a brush and mild soap solution to remove the crusts. All parts of all lesions should be treated daily, preferably two or three times each day.

Since ringworm is transmissible to humans, care should be taken when treating infected animals that contact is minimised and strict cleanliness is observed.

Ophthalmia.   Ophthalmia is also known as infectious ophthalmia, infectious keratitis, infectious bovine keratoconjunctivitis or pink eye. It is caused by the organism *Rickettsia conjunctivae* or the bacterium *Haemophilus bovis* in cattle and by a rickettsia in sheep. The organisms are transmitted by contact and by flying insects, especially flies. Because of this, the incidence of ophthalmia is highest in the months when the fly population is greatest.

The organism causes irritation of the membranes of the eyes which leads to watering, inflammation, and swelling of the membranes. After several days the cornea becomes cloudy. In some cases there may be some pus discharge from the eye. Untreated cases may result in blindness.

Treatment involves the use of antibiotic powders or

drops two to three times daily until a cure has been effected. The bovine forms will respond to penicillin, chloramphenicol, nitrofurazone and tetracycline. Both eyes should be treated because of the possibility of transmission from one eye to the other. In severe cases a subconjunctival injection of 1.5 ml penicillin and 0.5 ml cortisone preparation will often give a good response. This must be done by a veterinarian.

Although there seems to be some immunity in older animals, it has not been definitely shown. Vaccines have not proven very reliable in controlling ophthalmia. The best means of avoiding infection is to treat infected animals early and in isolation, avoid factors which might cause eye irritation such as dust or smoke, and control flies which may act as mechanical vectors.

5.5 WORMS

The worm burden

Tapeworms, flukes and roundworms cause debilitation and death in farm stock.

Tapeworms.   Adult tapeworms are long, white, flat worms, consisting of a series of segments gradually getting narrower toward the head which is often very tiny, round and provided with suckers and hooks. Adults are normally found in the small intestine or bile ducts of humans and other mammals. Larvae are found either in insects or farm animals in active muscle, heart, liver, lung and brain where they form cysts called measles, bladder worms or cysticerci. Eggs are found in the tail segments of the mature tapeworm. The eggs are very tiny and are normally not seen by visual observation.

Drenching will rid infected animals of adult tapeworms but there is no way of removing the larval forms from the animals.

Flukes.   Adult liver flukes are flat, greyish-yellow

worms 25 to 50 mm long. Adult rumen flukes are whitish,
8 mm long and conical in shape. Adult blood flukes are
about 6 mm long and roundish-flat in shape. Larvae are
extremely tiny and migrate in the tissues till they reach
the sites where they will mature. They cannot be seen
without a microscope but the traces of their migrations in
the liver are easily seen. Adult liver flukes live in the
bile ducts and their larvae migrate in the liver tissue.
Adult rumen flukes attach to the rumen wall and the larvae
live in the duodenum. Adult and larval blood flukes live
in the host blood vessels especially around the gut and
the bladder. Drenching is effective in eliminating adult
and larval liver and rumen flukes but the elimination of
blood flukes requires intravenous injections i.e. veterinary treatment.

Roundworms. Adult roundworms are round and white (unless coloured with blood they have ingested), have two
pointed ends and vary in length from 3 mm to 1400 mm. The
larvae are similar in structure but are microscopic in
size. Adults are most commonly found in the gut but sometimes are found in the lungs, trachea, kidneys, heart,
connective tissue and muscles. Larvae are found in most
tissues of the body. Most adult and larval stages found
in the gut, lungs, trachea and muscles can be eliminated
by drenching or injecting various worm remedies. There is
no way of eliminating worms from the kidney, connective
tissue and heart.

Elimination of the worm burden

If you suspect worm infestation you should send faeces
samples (preferably taken from the rectum rather than
collected from the ground) to the nearest veterinary diagnostic unit. Once the worm species have been identified,
purchase a worm remedy which is effective against the in-
fecting species. The dosages of most remedies are based
on the weight of the animal so you must in some manner

determine the weight (see 2.12, 2.13 and 2.14). Be accurate in your dosage; underdosage will be only partially effective and may lead to the development of resistant strains of worms whereas overdosing may kill the animal.

Oral administration (drenching). Animals may be drenched by means of a drenching gun, or a cool drink bottle or the outer shell of a cow horn. The drenching gun is a large, gun-shaped, automatic syringe which can be pre-set to give the same volume each time the handle is released. After the gun is filled, tip the animal's head up slightly (enough so that the liquid will run down the throat but not so much that it will be unable to swallow). Insert the nozzle of the gun into the back of the mouth (not down the throat) on top of the tongue. Expel the solution, making sure the animal swallows. Tickling the throat will sometimes cause swallowing in a reluctant animal. Freeing the tongue also aids swallowing. Be sure that the solution does not go into the larynx and thus to the lungs as this may result in the death of the animal. Once the animal has swallowed the remedy, the gun may be removed and the head released. Be sure to remove the gun from the animal's mouth before releasing the trigger or the solution will be re-aspirated into the gun.

Bottles used for drenching must be tall and narrow with a small opening and constructed from heavy glass. The former is to give a relatively slow flow to avoid the drench over-flowing into the lungs or out of the mouth. The latter avoids having the neck of the bottle bitten off. A Coca Cola or other cool drink bottle is ideal for this purpose. Bottles with horizontal ridges are especially useful as calibration lines can be placed in the bottoms of the ridges and not be washed off. Beer bottles are usually not strong enough.

When the remedy has been measured into the bottle, tip up the animal's head. Force open the mouth and insert the bottle at the side with the neck of the bottle about midway in the mouth. Do not insert the bottle so far back in

the mouth that the neck is between the molars or it may
be bitten off. Allow the solution to flow slowly into the
mouth. Be sure the animal swallows. Tickling or rubbing
the throat often will induce swallowing. If the animal
should cough or choke, release its head for about 30
seconds so it can clear its windpipe.

Using injectable worm remedies.   When large numbers of
cattle are to be treated for worms or where the animals
are large, it is often easier if one of the injectable
remedies is used. These are normally used with an automatic syringe which expels a premeasured volume each time
the handle is depressed. Manufacturers' directions for
use of the syringe should be followed with the injection
given intramuscularly into the rump or subcutaneously in
the neck. Not more than 20 cc should be injected into any
one site. If the required dose is larger than this, remove the needle and re-insert it into another location.

Worming programmes.   It is normal practice to try to
establish a pasture rotation programme to assist in worm
control. If such a programme is successful, it is usually
possible to establish a routine worming programme under
which all animals are dosed only two or three times per
year. If, for some reason, an adequate rotation cannot
be established it may be necessary to worm the stock much
more frequently. After each worming, the stock should be
moved to a clean pasture which has been rested at least
four weeks. Since larval tapeworms are not affected by
the worm remedies available, it is advisable to dose twice
at a two-week interval for tapeworm control. After
two weeks the larval forms will have matured and become
susceptible to the worm remedy. Control of liver flukes
is improved if livestock are prevented from drinking from
standing water supplies such as swamps and pans which are
infested with the secondary hosts (snails).

5.6   FEEDING CALVES

The objective in any calf rearing programme should be
rapidly growing, healthy calves at as low a cost as possible. When this objective is met, one should expect a
daily gain of 0.6 - 0.7 kg (for Friesians), providing a
calf at 12 weeks weighing 132 to 151 kg (assuming a birth
weight of 36 kg).

Calves may be raised in several ways including allowing them to suckle their dams or other cows, feeding
whole or skim milk from a bucket or feeding any one of a
number of milk substitutes from a bucket. Whichever system is used it is imperative that the calf receive its
mother's milk for the first three days. This milk, known
as colostrum, is rich in proteins and vitamins and contains antibodies against diseases to which the dam has
been subjected, and necessary for the calf's survival.
If the calf is allowed to suckle the cow for the colostrum, the cow should be milked out after the calf's appetite has been satisfied.

Raising calves by suckling

Allowing the calves to suckle the cow is probably the
easiest way to raise them, but it is expensive in terms
of the amount of milk taken in relation to that actually
required by the calf and it is also difficult to regulate
the amount of milk received by each calf. Although many
cows will allow only their own calves to suckle them, a
few will readily take on other calves. If one is fortunate enough to have a cow of this nature, it may be possible to raise as many as four calves at one time by
allowing them to suckle the same cow.

Bucket-feeding whole or skim milk

When milk is needed for sale or for feeding to the family,
it is usually more efficient to feed the calves whole or
skim milk from a bucket. This allows one to regulate

accurately the amount of milk received by each calf. The milk should be fed at body temperature for at least the first month. The feeding schedule in Table 13 is only a guide and must be carefully adjusted for each calf. As a rule of thumb, you should feed about 1 kg whole or skim milk or milk substitute solution per 10 kg body weight.

Table 13. Feeding routine for whole and skim milk.

Day	a.m.	p.m.
	Volume (litres)	
1-3	Allow calf to suckle cow for colostrum*	
4-6	3 whole	3 whole
7	3 whole	2 whole + 1 skim
8	2 whole + 1 skim	1 whole + 2 skim
9-10	1 whole + 2 skim	3 skim
11	3 skim	3 skim
12-14	3½ skim	3½ skim
15-21	4 skim	4 skim
22-84	5 skim	5 skim

* Strip out cow at completion of suckling.

Starting after one week, small amounts of concentrate should be offered to the calf. The amount fed should be gradually increased to 1.0 - 1.2 kg twice per day at 12 weeks. Good hay should be available at all times to provide bulk in the diet and to promote rumination. Any change-over from one type of feeding to another should be done gradually; preferably over a period of at least a week. Sudden changes will lead to digestive disturbances which can cause scouring leading to other, more serious conditions, and sometimes death. For the same reason the calves should be fed on a regular schedule.

Calves are not born with the ability to drink from a bucket so they have to be taught. To do this, stand with the calf's withers between your legs. Hold the bucket of milk in one hand and, after dipping the fingers of the other hand in the milk, place two fingers in the calf's mouth. As the calf sucks your fingers draw its head into the bucket and withdraw your fingers. The continued sucking by the calf will draw milk from the bucket. The calf will quickly learn that it can get milk by drinking from the bucket without the presence of your fingers. Do not leave your fingers in the calf's mouth any longer than necessary or it may require that the fingers be present before it will attempt to drink.

Bucket-feeding on milk substitute

There is a wide variety of milk substitutes available which vary in quality. How much you feed will depend on the quality of the substitute and the size of the calf. The change-over from whole milk to the milk substitute can proceed much as shown in Table 13 for changing from whole to skim milk. With any substitute you should carefully follow the manufacturers' instructions, be sure the substitute is thoroughly mixed in the water and feed at body temperature. If feeding at body temperature is not possible, at least be sure to always feed at the same temperature.

It is preferable to weigh the calves regularly, e.g. weekly, and feed on the basis of weight rather than age. This is because the calves will have different birth weights and, for a variety of reasons, will grow at different rates so feeding strictly on age can lead to serious errors in the amounts fed.

All calves should be carefully watched for the appearance of milk scours. If scouring develops, you should halve the milk intake for one or two feedings, then gradually raise the level to normal. Be sure to provide sufficient water to replace the liquid lost when the milk intake is restricted (see 5.4). Glucose may be added to the water to maintain the energy intake of the animal. Complex sugars such as sucrose should not be used because they are less readily absorbed.

After calves have been fed and their mouths are wet, they will tend to continue sucking anything they can get in their mouths including the ears, navels and udders of other calves. This is harmful to the sucking calf as it sucks in air and becomes bloated and may cause permanent damage to the other calf also; particularly to the udder. If the calves are given dry food such as grain after their milk the sucking will be greatly reduced. Some calves will continue to suckle regardless and should be isolated from the other calves. Young calves often do not recognise concentrates as something edible but can be encouraged if a handful of concentrate is sprinkled on the milk. When the calf empties the bucket and is consuming the last of the milk it will also eat the concentrate. In this manner it is often possible to start calves on concentrates at a very young age.

Young calves will often nibble at fine, high quality hay and this should always be provided where possible. Unless silage is of very high quality, it may cause scouring in calves under two months.

6 Sheep

6.1 SHEEP TYPE

Sheep may be kept for both wool or mutton or for mutton
alone. In either case, a solid and strongly built animal
with good body capacity and strong, well-muscled legs is
desirable. In addition, the sheep should have a broad,
strong back which does not sag. All the preceding are
characteristics which favour the animal's ability to feed,
produce and reproduce.

The sheep population of the world can be divided into
roughly two groups. The first group comprises those which
are historically of European origin. The second group
comprises the Eastern breeds of North Africa and Asia.
The majority of the former group produce a sizeable fleece
of relatively good quality wool. The second group are
kept basically for meat production, their coarse hairy
wool being used for carpet manufacture if used at all.

The breeds of European origin divide further into
sheep kept primarily for wool production and those kept
primarily for meat.

In most countries of the world where sheep are kept
primarily for wool production the sheep used are Merinos
or their derivatives such as the Rambouillet. Merino and
English Longwool derivatives such as the Corriedale are
also important wool producers, as are longwool crosses by
such sires as Lincoln Longwool, English Leicester and
Romney Marsh onto Merino.

Examples of European breeds where the emphasis is on

meat production are the English Southdown and its
derivatives such as the Suffolk and Dorset Down, together
with some continental breeds such as the Texel.

A number of European breeds are substantial milk producers, the milk being used for cheese making. In some
countries, especially in South East Europe, the sheep are
triple purpose animals and milk production is important.

A number of Asiatic breeds of sheep have the property
of storing fat on the rump or round the tail. This en-
ables them to withstand times of privation. Of the Eastern breeds, amongst the better known are the Blackheaded
Persian, the Dorper, Falani, Masai and Van Rooy. An Eastern breed with a world-wide reputation is the Karakul.
Astrakan fur is derived from the pelts of the very young
Karakul lambs.

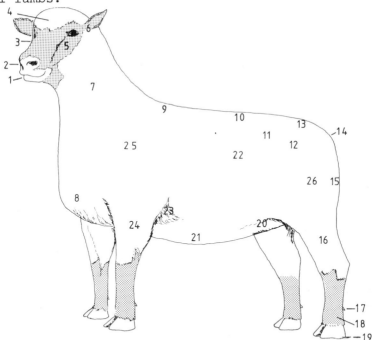

1. Mouth; 2. Nostril; 3. Face; 4. Forehead; 5. Eye; 6. Ear; 7. Neck;
8. Breast; 9. Top of shoulder; 10. Back; 11. Loin; 12. Hip; 13. Rump;
14. Dock; 15. Thigh; 16. Hind leg; 17. Dew claw; 18. Pastern;
19. Foot; 20. Hind flank; 21. Belly; 22. Ribs; 23. Fore flank;
24. Fore leg; 25. Shoulder; 26. Leg; 27. Twist.

Figure 38. Parts of a meat type lamb.

Figure 38 shows the conformation of a typical meat lamb. Ideally the animal should be straight backed, broad in the loin, full-fleshed above the hock and light in the fore-end. Breeding ewes should be of a size and weight appropriate to their breed, and special reference should be made to the soundness of feet, mouths and udders. Note that the names of some of the parts are different from those of cattle.

6.2   HANDLING SHEEP

Although sheep tend to be nervous and 'flighty' at times, they are generally docile and quite easy to handle. When sheep become excited or want to make a warning signal they stamp with the front foot. When moving sheep, their flocking instinct must be borne in mind and should be utilised to provide an easy flow of stock. Normally, if one or two leaders can be directed in the desired direction, the rest of the flock will follow quite readily. This also can work to your disadvantage if you are moving a flock of sheep and one or two escape. This generally means you will have to regather the entire flock.

Sheep should not be struck or prodded with sticks; nor should sheepdogs be permitted to bite them. If some inducement is necessary to get the sheep to move, rattling a plastic fertiliser sack near them is effective. Sheep must never be caught or held by the wool as this leads to bruising of the underlying muscle and weakening of the wool. Similarly sheep should never be caught by a leg because of the danger of breaking the bones. The proper method for catching a sheep is to approach it from the rear and grasp it round the neck with the left hand and the skin of the flank with the right.

For most operations the sheep is not laid on its side, but is placed on its rump with its back to the operator as shown in Figure 39(a). The procedure for casting a sheep is as follows:

Catch by the right flank with your right hand and move the sheep around so you are standing facing its left side with your feet about 30 cm apart between the sheep's front and hind legs.

Grasp the sheep under the throat with the left hand.

Lift and turn the sheep in one smooth motion to place it on its rump. Be sure the hind legs do not touch the ground while you are doing this. If one leg does touch the ground, the sheep will kick and you will be unable to place it properly. If this happens replace it on its feet and start over again.

Allow the sheep to lean back against your legs. If necessary, some pressure can be applied to the tops of the shoulders with the inside of your knees to confine the animal.

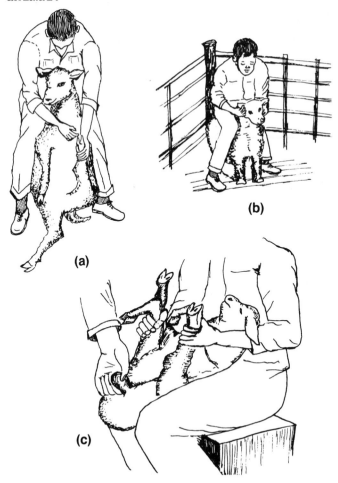

(a)

(b)

(c)

Figure 39. Restraining sheep.

Fat-tailed sheep should be allowed to tilt slightly to one side so they are not resting directly on their tails. In this position one man can hold the sheep and shear it, check its teeth or udder, vaccinate it, drench it or trim its feet.

Figure 39(b) shows one method of holding a sheep for drenching. The sheep should be caught and backed into a corner of the fence. If the operator then straddles the sheep, he can hold it, lift its head and deliver the remedy. Drenching can also be done with the sheep in the sitting position described above or standing when packed into a crush which is the fastest method when large numbers of sheep are treated at one time.

Castrating and docking lambs generally requires two people, one to hold the lamb and the other to perform the operation. The holder should be seated and hold the lamb on its back in his lap as shown in Fig. 39(c) with the lamb's right front and hind legs in the holder's right hand and the left front and hind legs in the left hand.

With sheep, as with any other class of livestock, one must always remember that careless handling can cause injury and that all movements must be positive and smooth, not hesitant and jerky. Careful handling of the sheep to avoid excitement will greatly facilitate work with the flock.

6.3   FOOT TRIMMING

The hooves of sheep grow continually, especially at the toe and along the sides. Sheep on natural grassland suffer more wear on their feet than those on irrigated pasture or other soft ground. As a result the former are less prone to the splitting and injury which often result from overgrown hooves. Whenever splitting and injury occur or when the sheep begin to walk improperly and before the rams are put into the breeding flock, the hooves should be trimmed. It is important that sheep are not allowed to walk on bad hooves as this may lead to crooked pasterns.

For trimming, the sheep should be positioned as shown in Fig. 39(a). Using a knife or hoof trimming shears, carefully trim away all excess hoof. Be careful not to cut too deeply or bleeding may result. Carefully check the bottoms and inner sides of the hooves for pockets in which dirt or manure may accumulate. These pockets should be opened and thoroughly cleaned. A hoof pick is a convenient instrument for cleaning and probing the hoof.

It is desirable that the hoof be trimmed so that the sheep is forced to walk on the front part of the hoof. This is accomplished by trimming the side wall flush with the bottom of the hoof and by shortening the tips of the toes. Be careful that the two tips are even.

After the feet are trimmed, the flock should be put through a footbath (see 2.17) of copper sulphate or formalin solution which will disinfect any open wounds which might have resulted from trimming too close.

6.4 SHEARING

Shearing means removing the fleece from a sheep in such a manner that the quality and thus the economic value of the fleece is maximised. The technique of shearing is very difficult to describe in words and is best taught by demonstration. The following list describes some of the factors which help to preserve fleece quality.

Shear in a clean place and do not allow foreign materials such as dung, straw, chaff or sand to get into the fleece.

Do not crowd sheep into small pens before shearing as the pens will rapidly become saturated with dung and urine resulting in dirty fleeces.

Dag all sheep prior to shearing, if necessary.

Remove the fleece from the sheep in one piece.

Shear close to the animal's body.

173

Avoid second cuts, i.e. do not go over the same area twice to try to smooth up the shearing job. This only serves to increase the number of short fibres in the fleece which will reduce its value.

After the fleece has been removed from the sheep it should be laid out flat with the flesh side down then folded with the belly wool in the centre. Fold in each side and roll the fleece from both ends. The fleece can then be tied using the wool off the legs or with paper twine. Sisal or hemp twine should never be used as fibres from the twine will become mixed with the fleece.

A pictorial description of the technique of shearing may be found on pages 199-204 of *Sheep Production* by Bundy and Diggins, Prentice-Hall, Englewood Cliffs, New Jersey, 1958.

6.5 LAMBING

With any sheep operation, but particularly in an operation which produces fat lambs, the profit to be made depends greatly on the size of the lamb crop at birth and at weaning. Although most ewes will lamb normally without assistance and most lambs will survive, good husbandry will increase the lamb crop.

Here are some of the husbandry factors of importance to lambing.

Use only healthy ewes and rams for breeding. Before breeding, check all the sheep for foot problems and broken mouths. Broken-mouthed ewes should be culled.

The ewes should be on a rising plane of nutrition at tupping and again before lambing. If ewes are too thin they may not be able to withstand the stress of lambing and will have a reduced milk supply for the lamb.

Exercise during the final weeks of pregnancy is important to maintain fitness.

Ewes should be vaccinated against enterotoxaemia at 6 to 8 and again 2 to 4 weeks prior to lambing (see 6.6).

At the second vaccination the ewes should be treated
for worms, if necessary, (see 6.9) and be shorn or
faced, dagged, and crutched, if necessary. Removal of
the wool keeps the ewe clean during lambing, reduces
the chances of infestation of the wool with parasite
eggs or disease organisms, makes it easier for the
lambs to find the teats and avoids woolball in the
lamb from swallowing wool when suckling.

A week before lambing is to begin, the ewes should be
placed in a clean pasture, which has been rested for
several weeks and in which lambing has not occurred
previously.

As ewes lamb, they must be carefully watched so those
which require it can be given assistance and to be
sure the lambs are accepted by the ewe and are allowed
to suckle. During the first few days the ewe knows
her lamb only by smell so care should be taken that
nothing interferes with this. After the first few
days the ewe also recognises her lamb by its voice.
A number of maternity or lambing pens should be constructed from hurdles or bales of hay or straw in
which animals requiring assistance can be housed.
These pens should be roofed and protected from wind
and rain to prevent chilling the newborn lambs. Pens
for yearling ewes should be 1.2 m x 1.2 m whereas
those for older ewes or those with twins should be at
least 1.2 m x 1.5 m. Once a ewe and lamb(s) have been
confined in the lambing pens, they should be kept
there for one or two days before being returned to the
lambing pasture. Any lamb which is chilled should be
placed in warm water (as hot as the hands can stand)
until signs of chilling disappear and then thoroughly
dried. N.B. Keep the head out.

Once the lambs are 2 days old they are quite hardy and
able to survive all but the most adverse weather and
should not be moved about unless absolutely necessary.
Management of young lambs is discussed in Section 6.6.

In summary, proper facilities and care are much more important than treatment in reducing losses of young lambs.
Most ewes will lamb normally without assistance but, if
assistance is needed, it is vital that the proper facilities are available.

6.6   LAMB MANAGEMENT IN GENERAL

The majority of lamb deaths occur in the first two weeks

175

of life. There are numerous causes of these early deaths
(as well as later ones) which may or may not be prevent-
able through proper management. A list of some of the
major causes of death and the management procedures which
help overcome them follows.

Injuries sustained during birth such as collection of
fluids in the head, liver rupture or fractured limbs
or ribs can cause death in the newborn lamb. These
often occur if the ewe becomes excited so she should
be left alone as much as possible during lambing. If
assistance is required, it must be provided in an extremely careful and gentle manner. Before putting
your hand into a ewe, wash your hands and arms and
trim your fingernails. Lubricate your hand and arm
with medicinal liquid paraffin or soap and water.

If the newborn lamb is not breathing, clean its mouth
and nostrils of mucus and membranes then blow gently
into the nostrils then gently slap it over the heart.

If the newborn lamb does not suckle or if, for some
reason, the ewe does not milk, the lamb will die of
starvation. If the ewe will not allow the lamb to
suckle or has no milk the shepherd must take corrective action. In many cases if there has been a difficult or disturbed birth, the ewe tends to be nervous
and will force the lamb away. She often will settle
down within several hours and, if closely confined in
a lambing pen with the lamb, will accept it. If not,
it may be necessary to tie the ewe's head and assist
the lamb to suckle. Very small or weak lambs may also
require help in suckling.

If the ewe has no milk, attempts should be made to
force the lamb on a ewe which has lost her lamb, or on
one which has sufficient milk to rear two lambs.
There are several ways of fostering lambs including
wrapping the foster lamb in the pelt of the dead lamb
or, if both are alive, soaking both thoroughly in salt
water to cover the smell of the lambs so the ewe does
not know which lamb is hers. Sometimes smearing
faeces from her own lamb on a ewe's nose will cause
her to accept a foster lamb. Carefully observe the
ewes when fostering lambs as they may try to butt
strange lambs and may injure them. Rearing orphan
lambs on a bottle is not advised since the death rate
is high and the problems encountered with these 'pet'
lambs are many. Bottle rearing also requires special
ewe milk substitute if it is to be successful.

If the lambs are exposed to long periods of cold, wet weather, many may succumb to pneumonia. The obvious method of prevention is to provide shelter during such weather. As the lambs grow older, they become less susceptible to exposure.

The faeces which result from colostrum are very sticky. This may stick the tail to the anus and dry so the lamb is unable to defaecate. Lambs should be watched and the faeces removed if it appears to be a problem.

If the lamb suffers an open wound or if the ewe is assisted during lambing without proper sanitary precautions being taken, the organism *Clostridium tetani* which causes tetanus may be introduced.

Prevention requires use of rigid sanitary procedures and, where tetanus is a problem, all ewes should be immunised with the toxoid with two injections 30 to 60 days apart. An annual booster should be given just prior to lambing.

Enterotoxaemia (pulpy kidney, overeating disease, apoplexy) is caused by a sudden release of toxins by bacteria of the genus *Clostridium* in the digestive tract as a result of stress. The stress may be a sudden change in management, feeding or the weather. The disease has high mortality and there is no cure. Prevention depends on good management i.e. make changes of feed slowly, minimise the worm burden in the lambs (see 6.9) be sure ample feed is available at all times and vaccinate the ewes 6 to 8 weeks and again 2 to 4 weeks prior to lambing. This confers immunity on the lambs for the first few weeks of life. All lambs should be vaccinated at 12 weeks of age. Do not de-worm lambs before vaccination unless the ewes were vaccinated.

Dogs and other predators can create havoc in a lamb flock. Precautions must be taken to keep such predators out of the flock. If, however, lambs are killed by predators, the carcass remains should be left as they are but injected with Toxaphene which will poison the predators should they return to the kill as they often do. Do not forget to tell your neighbours that you are putting out poison. Before putting out poison check that it is legal!

As pointed out previously, (Section 2.3) if any sort of selection programme is to be carried out, it is imperative that each individual animal be identified and records kept.

Sheep can be identified by ear tags (see 2.5), tattoos (see 2.6) or ear notches (see 2.7). For lamb production, records kept should include lambing results, birth weights and 12 and 18 week weights. Comparison of these weights provides an indication of the quality of the lambs and the mothering ability of the ewes. Twelve weeks is a convenient time for weighing as the lamb can also be vaccinated against enterotoxaemia at this time. The 18 week weighing should occur at or just prior to weaning. For wool type sheep, one should also keep records of the wool production of the lambs when shearlings.

A convenient programme for lamb management on an extensive sheep farm is shown in Table 14.

Table 14. Lamb management programme.

Age	Operation
1 day	Weigh and identify.
Before 7 days	Castrate (see 6.7) and dock (see 6.8).
12 weeks	Weigh and vaccinate against enterotoxaemia; deworm if necessary.
18 weeks	Weigh and wean; deworm if necessary.

Raising lambs can be a profitable enterprise, but requires careful, consistent attention to detail throughout lambing and the subsequent growth period.

6.7 LAMB MANAGEMENT - CASTRATION

Various methods of castrating calves have previously been described (see Section 5.2). The procedures using the elastrator and the Burdizzo are the same for lambs as with calves. Castrating lambs with a knife, however, is different in that the bottom third of the scrotum is removed. The testes are then pulled out to the white membranous sheath and the cords cut as high as possible. It is advisable to scrape the cord off as with calves to prevent haemorrhage. The wound should be treated with Stockholm tar or antiseptic. After the lambs are castrated they should be allowed to run on clean pasture.

As with the other classes of livestock, castration should be done as early as possible to minimise the growth check due to the stress of the operation. Some lamb producers use the elastrator band on the scrotum before the testes descend to make what are known as 'short-scrotum' lambs. Because the testes are kept within the body cavity no sperm are produced but the production of the male hormone by the testes will continue and accelerate the growth rate.

6.8 LAMB MANAGEMENT - DOCKING

Docking, also known as tailing, is the procedure of removing the tail of young lambs to help prevent fly strike. Docking may also make service by the ram easier and help avoid the build-up of parasites, especially ticks, under the tail. The tail may be removed using a knife, Burdizzo, hot chisel or elastrator rings.

When docking with the knife, the skin of the tail should be pushed toward the body and the tail severed at the root or one or two vertebrae down. The skin should then be allowed to slide down to cover the cut end. Although this is a one man procedure some bleeding may result so some sheepmen cauterise the wound with a hot iron. Painting the wound afterwards with Stockholm tar or antiseptic is advised.

The Burdizzo can only be used on small lambs. Using this procedure, the tail is clamped near the base and the end is cut off inside the Burdizzo with a knife. Little bleeding should occur but can be stopped by cautery if it does and the wound should be painted with Stockholm tar or antiseptic. This, too, is a one man procedure but has the disadvantage that the farmer has to own a Burdizzo.

When a hot chisel is used, the chisel is heated red hot in a fire. The tail is then placed on a board and the hot chisel used to cut it off. This method gives the least bleeding of any of the cutting methods of docking but does require two men. Healing may also be somewhat slower than with the other methods.

There are two methods of docking using the elastrator. The easiest is to place the rubber ring at the base of the tail where the two folds of skin join the tail and leave it until the tail drops off. This method, however, pro- vides great potential for infection. Alternatively, the ring may be left on the tail for 36 hours after which that portion of the tail below the ring (and the ring) are re- moved with a knife. No bleeding should result. The wound should be treated with Stockholm tar or antiseptic.

Whatever method is used, the lamb should be docked as early as possible to avoid a check in its growth and all possible care should be taken to avoid infection. Note also that some countries impose legal limits on the age at which docking can be carried out without a local anaes- thetic. In areas where tetanus is a problem, vaccination with antitetanus toxoid at or before the time of docking is advised.

6.9   DRENCHING SHEEP

The reasons and techniques for drenching calves have been discussed in Section 5.5. The techniques for sheep are similar. Sheep can be dosed in a crush, backed into a corner as shown in Fig. 39(b) or when placed on the tail (Fig. 39(a)). Use of the crush or fence corner is prob- ably faster than up-ending each sheep but the latter has an advantage in that as each sheep is drenched it can be examined also for tooth wear, foot problems, tick infest- ation, etc. thus combining several operations into one.

Several factors should be borne in mind when drenching sheep.

If dosing wool-type sheep with phenothiazine drenches, care must be taken that the drench does not come in contact with the wool as it will cause permanent staining.

Whenever possible, ewes in late pregnancy should not
be drenched as the stress of handling may be detrimental. Most of the newer drenching compounds are safe
to use on pregnant animals.

If using a carbon tetrachloride drench, be sure the
solution is released in the back of the throat as the
vapours, if inhaled, may be fatal to the sheep.

When drenching lambs destined for slaughter observe
the waiting period required between drenching and
slaughter.

In sheep heavily infested with parasites causing anaemia,
a characteristic swelling of the tissues of the lower jaw
is often seen. This condition is known as 'bottle jaw'
and is due to collection of fluids in these tissues.

As with calves, establishment of a good pasture rotation system is highly effective in the control of internal parasites in sheep. A special system often used for
ewes with young lambs is known as *forward grazing*. Under
this system a creep-type arrangement is provided in the
fence, so the lambs can graze forward into the next pasture which the sheep are to occupy. This allows the lambs
to graze clean pasture which is not being contaminated by
the adults and also allows them first chance at the new
grass.

7 Pigs

7.1 SWINE TYPE

Before the advent of vegetable oils for cooking, lard
(rendered pig fat) was in demand and pigs were kept which
laid down thick layers of backfat. These were known as
lard type pigs as opposed to the meat (pork or bacon) type
pigs which did not produce so much fat. When lard became
less saleable, pig breeders bred away from the excessively
fat animals and concentrated more on those animals which
would produce a meat animal quickly and efficiently.

At present, most pigs in the U.S. are marketed at 80
to 100 kg liveweight. In Britain, Europe and other areas,
meat pigs are marketed as porkers which are slaughtered at
45 to 55 kg liveweight, or baconers slaughtered at 80 to
100 kg liveweight. In most cases pork type pigs are shorter, more compact animals than baconers. Baconers tend to
be somewhat slower maturing and less inclined to excess
fat at the heavier weight.

Some of the older, more established breeds of pigs
are:

Berkshire	Landrace	Poland China
Chester White	Large Black	Tamworth
Duroc	Middle White	Wessex
Hampshire	Large White (Yorkshire)	Saddleback

Because of the pig's short generation interval, the development of new breeds is quite easy and is continually occurring. These new breeds and hybrid pigs, however, are based on the older, established breeds.

All pigs should have a strong back with a definite arch, and sturdy legs placed squarely under the body, as shown in Fig. 40. Note also that the neck is reasonably short and that the animal appears balanced.

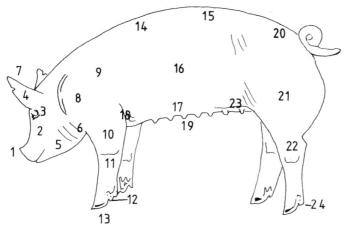

1. Snout; 2. Face; 3. Eye; 4. Ear; 5. Lower jaw; 6. Jowl; 7. Poll;
8. Neck; 9. Shoulder; 10. Foreleg; 11. Knee; 12. Pastern; 13. Foot-toe; 14. Back; 15. Loin; 16. Side; 17. Belly; 18. Fore flank;
19. Sheath; 20. Rump; 21. Ham; 22. Hock; 23. Rear flank; 24. Dew claw.

Figure 40. Parts of the pig.

7.2   MOVING AND HANDLING PIGS

Because of their odd shape, independent temperament and sharp teeth and tusks, pigs can be difficult to handle. Figures 41, 42 and 43 depict several techniques in handling pigs.

Figure 41(a) shows a pig snare. This device is used for catching and tying hogs. The snout is caught in the loop which is then pulled tight.

Figure 41(b) shows how to tie a pig. Using a snare or a loop of rope over the snout, a pig can be tied to a post or other solid object. Try to get the loop as far back in the mouth as possible. If the pig refuses to open its mouth to accept the loop, try to make it squeal or get it angry. Either of these will cause it to open its mouth.

Figure 41(c) shows a method of casting a pig. A medium sized pig can be cast using this technique.

Figure 41(d) shows how to cast a large pig. If the pig is over 70 kg, casting by means of a rope will probably be necessary. Sometimes sows will lie down if the udder is stroked.

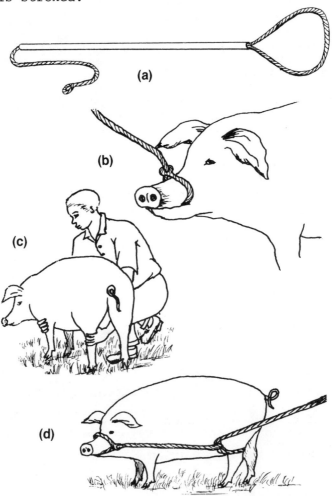

Figure 41. Catching and casting pigs.

Figure 42(a) shows how to guide a pig using a bucket. Pigs are very difficult to drive especially if being moved one at a time. Placing a bucket over the pig's head

causes it to back away from the bucket. The man at the
rear guides the pig with its tail, while the man at the
front keeps the bucket over its eyes. This is an especially useful technique for leading into a lorry.

Figure 42(b) demonstrates how to guide a pig using
panels. For movement over smooth ground, restricting the
side vision of the pig with a panel (hurdle) is often sufficient to guide it in the desired direction.

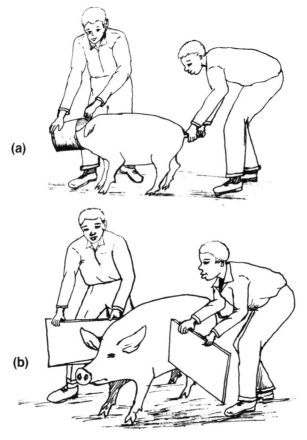

Figure 42. Moving pigs.

Figure 43(a) shows how to hold a piglet. This will be
necessary for cutting wolf teeth (see 7.6), ear notching
(see 2.7) and weighing. It is important that the piglet
feels it is securely supported and fairly tightly restrained.

Figure 43(b) shows holding a pig for vaccination.

This obviously is only suitable with pigs up to 50 kg. The pig should be held with its belly forward with its head off the ground.

Figure 43(c) shows how to hold a pig for castration. This is the same as 43(b), except the pig is held with its back forward.

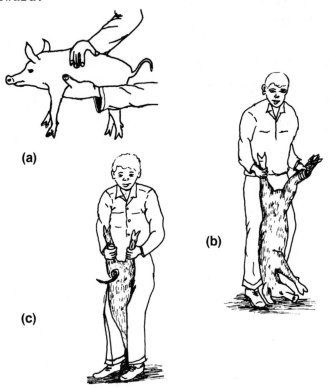

(a)

(b)

(c)

Figure 43. Restraining small pigs.

It is easier to move groups of pigs than individuals. Sticks should not be used with pigs any more than with other classes of stock. Sticks are practically useless anyway with pigs since they have thick skins with a thick fat under layer, and are relatively insensitive. A slap with the hand is just as effective. If many hogs are to be moved, a slapper can be made with a piece of heavy canvas tacked to a handle. This will be as effective as a stick and will not injure the pigs.

In all operations involving moving or handling pigs,

it is important to move quietly and deliberately so that the pigs do not get excited: excited pigs are difficult, if not impossible, to handle.

7.3   FARROWING

In a pig production enterprise, which depends on the sale of pigs for income, farrowing can be a critical time since the number of pigs per litter can spell the difference between profit and loss. A number of factors affect the number of live pigs weaned, including the facilities, management of the sow before and after farrowing, and the management and handling of the piglets.

The facilities

The facilities for farrowing need not be extensive, but they must provide a dry, draught-free place for farrowing to occur. There will be less danger of the sow savaging the piglets if the farrowing pen is visually isolated from other pigs.

Because of the extreme susceptibility of the piglets to disease, the farrowing pen must be thoroughly cleaned prior to each use. Thus, the pen should be designed to facilitate washing and disinfecting (see 1.9). In addition provision must be made so the sow does not lie on the piglets. This can be accomplished by putting the sow in a crate with a space along the bottom through which the piglets can escape or by installing a pipe rail about 230 mm from the floor and 230 mm away from the wall all the way around the pen. This is effective because large pigs often lie down by sliding down the wall. The rail protects any piglets which may be caught along the wall.

If there is any chance of chilling the piglets, bedding may be provided in the farrowing pen. This may be chopped hay, straw or maize stalks. Care should be taken not to use too much or the piglets may be smothered.

Pre-farrowing management of the sow

Four to five days before the farrowing, the sow should be thoroughly washed with water and soap or shampoo and disinfectant to remove dirt, germs, external parasites and eggs of internal parasites. The sow should then be placed in the clean farrowing quarters to allow time for it to become acclimatised to the new surroundings before farrowing occurs.

Two or three days before farrowing the feed level should be reduced and a laxative ration containing wheat bran provided.

Signs of imminent farrowing:

1. The temperature may rise a few days before farrowing but this tends to be a rather unreliable sign as the temperature is quite variable.

2. About 24 hours prior to farrowing, the sows become extremely restless.

3. Milk in profuse amounts can be expressed from the teats 16 to 8 hours prior to farrowing.

4. At least 5 hours before farrowing the sows will begin to arrange a nest if bedding is available.

5. About 3 hours before farrowing begins, abdominal contractions may be noticed.

6. Within 2 hours of farrowing, blood or bloody fluid will be expelled from the vulva.

Farrowing and post-farrowing management of the sow

If possible, it is probably best to observe the sow during farrowing in case it becomes necessary to provide assistance to the sow or the piglets. Possible instances in which assistance might be necessary include:

Preventing the sow from savaging the piglets. Some sows, and particularly some gilts, are extremely vicious towards their piglets until the actual process of farrowing is completed. If the piglets are removed until the sow settles down, they will normally be accepted without trouble.

Extrication of piglets born enveloped in afterbirth.

Preventing the sow from crushing piglets by lying on them.

Assisting weak piglets to nurse.

If farrowing occurs in a pen where there might be danger of infection, the navels of the piglets should be dipped in a 15% tincture of iodine solution.

Since, in spite of the number of things which might go wrong, most sows, especially the older ones, will farrow with no problems, the sow should not be disturbed during farrowing unless absolutely necessary.

After farrowing the sow should be left undisturbed as much as possible. Feed levels should be increased gradually. When the sow is on full feed it should receive 1 kg plus 0.5 kg for each piglet in the litter.

The management procedures required at farrowing are not difficult, nor do they require a great deal of labour. They can however, mean the difference between profit or loss in your swine operation so should be carefully followed at each farrowing.

7.4 PIGLET MANAGEMENT

When the piglets are born, they must be provided with a warm nest out of reach of the sow. In addition they require iron supplementation (see 7.5), the wolf teeth should be clipped (see 7.6), and they should be identified (see 2.7). All these operations necessitate handling the piglets which will squeal whether or not they are being hurt. Since a squealing piglet may upset the sow and cause her to harm the rest of the litter, it is best to perform these operations out of her hearing, if possible, and to separate her from the entire litter while you are working with the piglets.

Piglets are very inquisitive and will begin to nibble at feed as early as seven days of age if the feed is available and they will consume significant amounts by two or three weeks. The milk production of the sow tends to fall after about three weeks so supplementary (creep) feeding for the piglets is important, especially with very large litters or if the sow is in poor condition. It has been shown that creep-fed piglets may grow 20% faster than non-creep fed piglets. If the piglets are accustomed to eating solid food, it also reduces the check in growth at weaning.

The feed should be provided to the piglets in a creep, i.e. an area to which the piglets but not the sow have accesss. It is often convenient to make the creep the place where extra heat (if necessary) is provided for the piglets. With this set-up, only one place where the piglets are protected from the sow need be provided and the piglets do not have to be trained to enter for food as they are already accustomed to entering the area for warmth. This will tend to encourage the early utilisation of the creep.

Piglets may be weaned at any age from one to eight weeks although the last is probably most common except in commercial herds where very high levels of sow productivity are desired. By eight weeks the sows are producing very little milk and the piglets will have been nibbling at solid feed for some time.

Unless absolutely impossible, weaning should be carried out by moving the sow to new quarters rather than moving the piglets. The sow should be moved to a pen where she is out of contact with the piglets so continued visual or auditory stimulation will not upset the piglets or the sow. Separation from the sow and the lack of milk in the diet are both stresses and addition of another stress - becoming used to new quarters - will provide even more check to the piglets. After a further four weeks in the pen on solid food the piglets may be moved to fattening pens if desired.

About a week before weaning is to occur, the amount of feed given to the sow should be reduced by about one third. This will help to ensure that she dries off promptly when she is moved away from the piglets. After weaning, observe the udder to be sure that drying off has occurred before increasing the feed to normal levels. When reducing feed for the sow, be sure that the feed available for the piglets is not also reduced. This will be no problem if the piglets are being creep-fed.

7.5 PIGLET MANAGEMENT - IRON

Piglets have a very fast growth rate and will multiply their birth-weight by five times by the time they are 20 days old. Because of the rapid growth, there is also a rapid increase in blood volume. Consequently, haemoglobin, which has iron as its basis, is in increasing demand. It has been estimated that a piglet requires 7 mg of iron per day during the first three weeks of its life. The birth reserve of iron is approximately 45 mg and the iron provided from the milk is about 1 mg per day. Thus, without iron supplementation, the iron supply will be used up within eight days and the piglet will show symptoms of deficiency (anaemia). These symptoms include weakness, rough hair, wrinkling of the skin over the shoulders, neck and legs, lack of vigour, pale mucous membranes, white diarrhoea, lowered resistance to infection and sudden death.

It is obvious from the above that piglets need iron supplementation. This can be supplied from several sources:

If the pigs are maintained on pasture, the piglets will obtain their requirements from rooting in the soil.

Clods of soil may be placed in the pen or creep. Care must be taken that the soil provided is not contaminated by worm eggs or larvae.

An oral dose of iron paste may be given on the third and tenth days.

A solution of ferrous sulphate may be painted on the sow's udder each day. This is mixed by dissolving 500 gm ferrous sulphate in one litre of hot water.

Iron may be injected in a single dose of solution containing at least 100 mg iron. This probably is the best method as you can be certain that the piglet has received the iron. On large scale enterprises this is the only effective way of providing iron.

Iron fumarate in a cereal base may be scattered on the floor of the creep from the third to the twenty-seventh days of life.

No matter what form of supplementation is used, it is important that the piglet receive some form of iron early in its life.

7.6   PIGLET MANAGEMENT - WOLF TEETH

Piglets are born with sharp, well-developed, temporary corner and canine teeth. These are variously known as wolf teeth, eye teeth or needle teeth. Since these teeth can inflict damage on the sow's udder and to the other piglets, it is advisable that they be clipped at birth.

Cutting these teeth is a simple job requiring only a strong pair of sharp clippers. Hold the piglet as shown in Fig. 43(a). Lift the lip and cut the teeth at the level of the gum. Be sure you do not twist the clippers and break off the tooth below the gum, as this may lead to infection. Also, remember that there are eight teeth to cut, two top and two bottom on each side.

This, and any other operation, which might cause the piglet to squeal, should be done out of the hearing of the sow. The squealing of one of her piglets may upset her to the extent that she will savage the rest of the litter. In fact, it is a good idea to isolate the sow from the entire litter before carrying out the operation described in this section. Each piglet can then be treated and the litter returned to the sow when all have been done.

7.7 PIGLET MANAGEMENT – CASTRATION

In order to prevent undue stress to the piglet and to reduce the effort of restraint, it is best to castrate as
early as possible. Any time between 10 and 20 days of age
is suitable. If done earlier than this, it may be difficult to find some testicles and if done later there may be
undue bleeding. Castration should be done as a single
operation and not combined with vaccination, weaning etc.,
or the pigs may suffer a very severe check.

Because of the placement of the testicles in pigs, it
is not possible to castrate them with a Burdizzo or an
elastrator so a knife must be used. The procedure is as
follows:

Catch the pig and hold as shown in Fig. 43(c).

Examine the scrotal area to be sure it is normal. If
there is a large swelling which feels soft the pig may
have a scrotal hernia. Pigs with this or any abnor-
mality of the scrotum must be castrated by a veterinarian.

Thoroughly clean the scrotal area with methylated
spirits or disinfectant (see 1.9).

Grasp one testicle and push it upward to tighten the
scrotal skin.

Using the special castration knife or new razor blade
make an incision down the length of one testicle. Cut
only through the skin and white membrane.

Pull the testicle through the opening, twist twice and
scrape through the cord.

Repeat for second testicle.

If available, some antiseptic powder such as sulphanilamide or a tube of intramammary penicillin packed
into the wound may help prevent infection.

Cleanliness throughout the operation is essential, if infection is to be prevented. After castrating it is best to allow the piglets to run on clean pasture or in a clean pen as the exercise will promote drainage, thus healing, of the wounds (see 2.19).

7.8 PIG VICES

Pigs are prone to several vices including tail biting, cannibalism, over-aggressiveness, abnormal sexual behaviour and rooting.

Tail biting is probably the most common and serious of these vices. It often happens that the tail biting is carried on by only one pig in the pen and if this pig is removed the biting will stop. Another method of preventing tail biting is to remove all the tails at or soon after birth. This may be done in a similar fashion to the methods used for docking sheep. It is important that tail biting be stopped if it occurs as it is a contagious habit which the other pigs may pick up. It may also lead to infection or death.

Cannibalism and over-aggressiveness are most commonly seen in sows with new litters. Cannibalism may occur at farrowing if the sow is upset or disturbed (see 7.3) although some sows, especially young ones, will attack the litter whether disturbed or not. Over-aggressiveness is a means of protecting the litter and usually disappears when the litter gets a few days or a week old. Such behaviour may also be seen with pigs which are deaf or blind or both. Some pigs, however, do not like people and are aggressive all the time. If there are people about and danger of them being injured, such animals should be culled. If there is little danger to people, such pigs can be kept as long as the workers remember that the animals may be dangerous.

Abnormal sexual behaviour may occur if groups of pigs of the same age and sex are kept together in large groups for long periods. This vice usually disappears if the

pigs are allowed access to the other sex. Sexually
experienced boars may masturbate if they are used infrequently. This does no harm to the boar so no corrective
action is necessary.

Rooting is an expression of the normal wild behaviour
of pigs. Before they were domesticated, pigs were basically forest animals and obtained their food by digging in
the ground (rooting) with their snouts. If pigs are on
pasture or even in concrete pens, they will root if bored.
If it is important that the ground cover not be disturbed,
rooting can be prevented by placing one or more wire rings
in the top edge of the nose (in the dorsal margin of the
rostrum suis).

7.9 PIG DISEASES

Diseases in pigs may be caused by viruses, bacteria or
protozoa. They are also susceptible to internal parasites
such as roundworms and tapeworms and to attack by external
parasites such as lice and mites. Sick pigs do not produce at a maximum rate and many of the diseases to which
pigs are susceptible are highly contagious so it is important to identify, isolate and treat sick pigs as early as
possible. Careful observation will often allow the pig-
man to identify animals which are sick or have parasite
infestations. As with any other class of livestock, it is
important to know what a normal pig looks like. Any de-
viation from the normal appearance or behaviour should be
investigated as such deviations are often the first signs
of disease. Some of the symptoms that may be seen are
listed below.

A pig not eating or lacking interest in food.

A pig which is breathing rapidly.

A pig with harsh, rasping breath or a cough.

White pigs with a flushed (red) skin.

Pigs with general listless attitudes including drooping ears, dull eyes, dry snout, limp tail and lack of
lustre of hair and skin.

Production of loose, stringy or bloody dung.

Pigs with dry scaly skin or which spend excessive
amounts of time rubbing.

Pigs with blotches on the skin.

Any of the above may be signs of sickness and any pig ex-
hibiting one or more of these symptoms should be carefully
examined to determine the cause. Early treatment is often
more useful and efficient and less expensive than later
treatment.

The viral and bacterial infections of pigs are often
highly contagious and any pig with an elevated temperature
or symptoms of a contagious disease endemic in your area
should be examined by a veterinarian.

Pigs are susceptible to infestation by roundworms and
tapeworms. The roundworms may cause unthriftiness, loss
of weight and abnormal respiration. Tapeworms are especially important as the stage which is found in pigs is a
part of the lifecycle of the tapeworm which infests humans.
In areas where tapeworms are endemic, pigs normally contract them by contact with human faeces. Since infestation in the pig carcass may lead to condemnation of part
or all of the carcass, proper disposal of human faeces is
important.

Pigs with worms can be treated with a variety of compounds of which the piperazine compounds provided in the
feed or water are probably most common.

Lice and mites on the skin cause severe itching and
the pigs will be seen rubbing fences and the sides of the
pens. Severe cases of mite infestation (mange) may show
as scabby looking areas on the skin. Treatment of both of
these conditions is by use of chemicals such as those used
for dipping cattle. Normally these products will contain
information on the label for the concentrations to use on

other classes of livestock. Once the material has been
mixed it is poured over the pig, making sure that all
parts are thoroughly wetted. Repeated treatments may be
required to clear up any infestation. If pigs do become
infested, their quarters should also be treated as the
pigs may be reinfested from unclean buildings or equipment.

7.10 HOUSING AND FACILITIES

There is no standard design of pig housing which will fit
all conditions. The design will depend upon the climate,
the particular pig enterprise (i.e. breeding or fattening)
and materials available locally.

In countries where there is a cold winter period and
a hot summer, the housing should allow for free ventilation during the hot months and for insulation as well as
ventilation during the cold months. One way of achieving
both is to use hinged wooden flaps along the top side of
the wall which can be opened or closed according to the
prevailing temperature. Air inlets for ventilation during
cold weather should be set 1.5 to 1.8 m above ground level
so that the cold air does not blow directly onto the pigs.
The outlet should be a central chimney large enough to
allow 4000 mm^2 of outlet per pig housed.

It is desirable to insulate the floor so that the pigs
do not get chilled from lying on cold concrete. The insulation can be provided by trapping air underneath the surface of the concrete. Various cheap materials are available which will accomplish this. Empty bottles or egg
trays can be used - the bottles are laid on their sides in
rows, the egg trays laid flat. The floor is then poured
with an adequate skimming of concrete 10-15 mm thick over
the insulating materials.

In areas where temperatures are generally higher, a
more open type of piggery can be used. The number of pigs
per pen is largely dictated by the trough space - the
space requirements per pig were shown in Table 2 in

Chapter 1. For fattening pigs, the required floor space
is nearly one square metre per pig housed. A breeding
sow should have a pen of 3.0 x 3.6 m including the dunging
passage. Farrowing rails should run around the pen: for
this use 25 mm galvanised pipe set 230 mm from the floor
and 230 mm from the walls. Each sow pen should have an
area to which only the piglets have access. This is known
as a creep and fresh food and water should always be
available in this area. This allows for a very important
aspect of pig feeding known as creep feeding (see 1.3).

When designing a piggery, bear in mind that it should
be easy to feed the pigs and to clean out the pens. The
feed trough and creep should have easy access: the floor
should be gently sloped towards the door so that water and
urine can flow away easily.

Where possible, a dunging area should be attached to
each pen. Contrary to popular belief, the pig is a clean
animal and will not normally dung or urinate in its sleeping or feeding quarters. Drinking water should be provided in the dunging area which should be in the daylight
if possible because pigs like to dung in the light. Pigs
can easily be trained to use a dunging area if the rest of
the pen is kept scrupulously clean and some dung is left
in the dunging area. Once the pigs are trained, cleaning
is relatively easy as all the dung, urine, and waste
drinking water will be confined to one place.

8 Poultry

The word poultry is generally taken to mean chickens,
ducks, geese and turkeys. In this section we are con-
cerned only with chickens (which are also known as fowls).
Chickens have two primary products; meat and eggs. In the
larger production units the birds are generally kept for
one or the other and are known as broilers and layers,
respectively. The small farmer, however, usually wants a
bird which will lay eggs for a year or two and then be
suitable for consumption.

There are several ways of classifying chickens. One
is to use the areas of origin as the major category i.e.
American, Mediterranean, English and Asiatic. These are
known as classes. The breakdown within classes is into
breeds which are differentiated on the basis of body
shape and size. Varieties within breeds are distinguished
by plumage colour and comb shape.

Egg type

Chickens in the Mediterranean class are primarily for egg
production. They have large combs, mature early, are nervous but relatively non-broody and produce white shelled
eggs. Breeds include:

Ancona	Leghorn
Andulasian	Minorca

American class egg producers yield brown eggs, have yellow skin, non-feathered shanks and red earlobes. Breeds include:

New Hampshire	Rhode Island Red
Plymouth Rock	Wyandotte

Meat type

The English class are primarily meat producers and lay mostly brown-shelled eggs. They all have white skin (except the Cornish). Breeds include:

Australorp	Dorking	Sussex
Cornish	Orpington	

The Asiatic class are also primarily meat producers. They are characterised by having feathered shanks and toes and producing brown-shelled eggs. Breeds include:

Brahma	Cochin	Langshan

It should be noted that the Plymouth Rock and New Hampshire breeds of the American class have been listed with the egg producers but they also produce very good meat carcasses so may be classified also as dual purpose. The Australorp, on the other hand, has been listed as a meat breed but this breed is also fairly efficient at egg production so could also be listed as a dual purpose breed.

Another means of classification is to divide the birds by body weight into light and heavy breeds. The light breeds generally lay more eggs but they only lay for one season and the eggs are smaller than those of the heavy breeds. In addition, they tend to be more prone to cannibalism and are more nervous than the heavy breeds. With a mature weight of less than 3 kg they provide a less meaty carcass. The heavy breeds have a mature weight of over 3 kg and are meatier than the light breeds. They are less excitable and will lay for several seasons but lay less eggs and are more prone to broodiness than the light breeds. Representative breeds of the two types are listed in Table 15.

Table 15. *Light and heavy breeds.*

Light	Heavy
Ancona	Australorp
Andulasian	Dorking
Minorca	New Hampshire
White Leghorn	Orpington
	Plymouth Rock
	Rhode Island Red
	Wyandotte

In addition to these breeds, there are also fowls known as hybrids which are special crosses combining the best characteristics of the parent breeds. These birds are bred for specific purposes e.g. the Hyline and Shaver which are specifically bred for egg production or the Indian River and Starbro which are specifically for meat production. Hybrids must be produced by special breeders and must not themselves be used for breeding as the high production characteristics of the parents are not passed to the offspring. Usually hybrid layers are purchased from hatcheries at one day old and brooded and maintained until the point-of-lay. These birds are generally kept for only one laying season. After they have been culled they will be sold and new birds purchased to replace them. Similarly, broilers are purchased as day-old chicks and raised to 9 to 12 weeks when they are slaughtered. This type of production requires very careful management and a fairly large capital outlay and therefore is usually not satisfactory for smallhold farmers.

Figure 44 shows a typical fowl and the names of the various parts of the bird. A good quality bird should have a relatively broad, level back without a sharp break between the back and the tail. The breast should be full

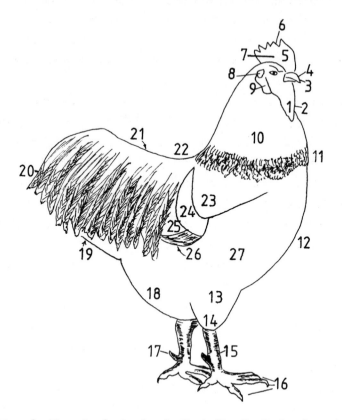

1. Wattle; 2. Throat; 3. Beak; 4. Nostril; 5. Comb; 6. Points;
7. Blade; 8. Ear; 9. Ear lobe; 10. Hackle; 11. Neck; 12. Breast;
13. Thigh; 14. Hock; 15. Shank; 16. Toes; 17. Spur; 18. Fluff;
19. Saddle feathers; 20. Sickles; 21. Saddle; 22. Back; 23. Wing bow;
24. Wing bar; 25. Secondaries; 26. Primaries; 27. Body.

Figure 44. Parts of the fowl.

and the abdomen should be expanded, soft and free from
fat. The bird should be alert with bright, prominent eyes
and should have no obvious abnormalities. It should walk
gracefully with its head up without evidence of lameness,
sore feet or crooked legs. Rough feathers are not necessarily a sign of a poor bird as birds which are laying
heavily will often have dirty, ragged feathers (see 8.5).

8.2 HANDLING POULTRY

Poultry are probably the most nervous and easily excitable

species of domestic livestock. Since with poultry, as with other species, production is improved if they are excited as little as possible, a few simple techniques and practices when working in a poultry house and handling poultry will increase your profits.

When entering a poultry house, announce yourself. Talk as you approach the house and knock on the door before opening it. This serves as a warning to the birds that something is about to happen so they will not be startled. When you enter the house, keep your movements slow and deliberate. If you are carrying a large object such as a bucket, keep it as low as possible so the birds will recognise it and not be frightened.

If birds must be caught, the best time is in the evening after they have gone to roost. If you must catch a bird during the day, keep your hand low and grab the bird quickly. Do not chase the birds around the poultry house. If possible, you should use a catching crate which is a structure with a wood frame covered with wire netting. This allows the catching of 25-30 birds at a time. You may also use a piece of netting attached to the house wall about two metres from the corner. The birds should be driven into the corner and the netting drawn around them. For catching individual birds, a catching hook which slips over the bird's leg is useful. When catching birds be careful that they do not pile up in the corners and suffocate those on the bottom of the pile. This can be a serious problem on hot days.

Once a bird has been caught, the preferred way to hold it is in a sitting position with its breast on the palm of your hand. Its head should be toward you so you do not push against the natural lay of the feathers when positioning your hand. If one leg is held between your index and middle fingers and the other between your ring and little fingers the bird will be securely held yet comfortable. This method is preferable to holding it by the legs with the head down as a bird in that position may vomit and choke. If the bird is held by the wings, bones may be broken.

If an intramuscular injection into the breast is necessary, holding the bird on its back with the wings held between your fingers in a manner similar to the method described above will provide solid support for the bird while it is firmly held in one hand. As with other species, a chicken which is firmly and securely held will not struggle and stands less chance of broken bones or other injuries.

To get a bird to lie quietly for a few minutes - perhaps for weighing on a small scale - place its head beneath one wing and gently rock the bird up and down three or four times. If placed on its side on the scale platform it will remain quietly for several minutes while a reading is made.

Your routine for feeding, collecting eggs, etc. should be the same each day. Poultry are creatures of habit and do best with a consistent schedule. If you move birds to another house, it is best to do so late in the evening. They can then be removed from the roosts in their old house and placed on the roost in the new house. If they are moved quietly without awakening them, they will continue to sleep on the new roosts and will have the entire next day to become accustomed to the house and its fittings before they need to find the roosts the next evening.

Proper handling of poultry involves a thorough knowledge of the things which will frighten and upset them and how to move and work among the birds in such a manner as to minimise these factors.

8.3 MANAGEMENT OF CHICKS

The management of chicks for the first five or six weeks is similar whether they are to be used in the laying flock or as broilers. The primary requirements of these chicks are that they are provided with a suitable environment, receive the proper diet and learn to eat and drink while being protected from disease.

Because these requirements are fairly stringent, especially with regard to temperature, the chicks are normally reared in specially constructed buildings known as brooder houses. In all but the largest, most specialised units, heat will be provided for the chicks under small structures known as hovers or brooders. These are small circular or rectangular roof-like units under which heat from electric, paraffin (kerosene) or gas heaters is provided. The hover generally will not exceed about 1.5 m in its longest dimension so the workers can easily reach under it when it is hung 200-250 mm from the floor. In the large poultry units, chicks may be brooded in cages and the entire building heated to the temperature required by the chicks. This is an expensive method unless many chicks are being handled and will not be discussed here.

When setting up hovers and brooding equipment the space requirements for the birds must be considered. Each chick should be allowed 4,500 to 6,500 mm^2 floor space under the hover, and 46000 mm^2 overall. At six to ten weeks of age the birds will need 0.10 m^2 floor space. If laying birds are to be kept in the same unit until point-of-lay, each bird will require 0.14 to 0.19 m^2 for the lighter breeds such as Leghorn whereas the heavy breed replacements will need 0.23 to 0.28 m^2. It is important that the space requirements be carefully met as too little space will result in the birds having problems finding feed and water which may lead to feather picking and cannibalism. If there is too much space the birds may become bored, which may also lead to the development of vices.

Before the chicks arrive, the brooder house and the equipment to be used must be prepared. The first step is to thoroughly clean and disinfect the building and equipment, according to the following procedure:

The ceiling and upper portion of the walls should be brushed and whitewashed if necessary.

The floors and lower walls should be scrubbed with hot water.

All equipment should be cleaned.

Any repairs required on the buildings and equipment should be carried out.

The interior of the house and all equipment should be thoroughly disinfected (see 1.9).

After the house is cleaned and disinfected, the heating equipment should be set up and tested to be sure it maintains the proper temperature. Within a day or so prior to the arrival of the chicks, sun-sterilised litter should be spread 100 to 125 mm deep over the floor. This litter serves as an insulator and serves to absorb moisture from the droppings. In order to maintain its usefulness after the chicks arrive, the litter surface should be stirred two or three times a week and fresh material added until a final depth of 180 to 225 mm has been reached. Suitable materials for litter include sawdust, wood shavings, broken maize cobs, peanut (groundnut) shells, etc.

Hoverguards

When the litter is in place, a small cardboard, wood or metal barrier should be placed around the hover 0.75 to 1.00 m out from the edge. This barrier should be 380 to 450 mm high. It is known as a hoverguard and serves to keep the chicks under or near the hover. A small light under the hover for the first few days and nights will make it easier for the chicks to find the heat source. After the first week or 10 days the hoverguard should be removed to allow the chicks free access to the entire brooder house. No corners should be left in the hoverguard or building. Corners of the room can be rounded with triangular pieces of board leaned into the corners.

Water

Many small water fountains should be used during the first
few days. These should provide at least 10 mm of drinking
space per chick under the hover and should be widely scattered through the area used by the chicks. The scattering
is necessary so that the chicks will find the water and
learn to drink. After the first week the water fountains
should be placed on a firm base, such as a piece of wood
or a brick, to prevent them from tipping over and wetting
the litter. The water in the fountains must be kept absolutely clean and the fountains cleaned at least once each
day. As the chicks get older and more accustomed to
drinking, less fountains need be used but larger quantities of water will be required, thus necessitating larger
fountains. Birds above two weeks of age will require
15 mm drinking space. The waterers should be distributed
so no chick has to move more than 4.5 m to water.

Feed

The chicks should be fed chick mash in very shallow trays
for one to two weeks. After two weeks the shallow trays
can be replaced by deeper ones. During the first few
days while the chicks are learning to eat the mash, the
trays should be placed on newspaper and filled to over-
flowing. This helps the chicks to find the mash while the
newspaper discourages them from eating litter. Once the
chicks have learned to eat, the trays should be filled
only two-thirds full or less. The space requirement at
the feeders is 25 mm per bird during the first two weeks,
50 mm during the third to sixth weeks and 75 mm for birds
of seven weeks and older. Feed trays should have wire
guards or a roller over them to keep the birds out of the
feed and from perching on the feeders.

Temperature

As previously mentioned, the temperature of the brooder is

critical to successful chick rearing. Day-old chicks require a temperature of about 32-35°C. This should be reduced by 2 to 3°C each week until about the sixth week or
until a temperature of 21-24°C is reached. Rather than
try to use a thermometer to guess the chicks' requirements
for heat, it is best to watch the chicks for some very obvious signs:

> If the chicks are found clustering around the walls or
> hover guard and are breathing heavily, it is too hot.

> If the chicks are piled up under the brooder, it is
> too cold.

> If the chicks are crowded in a corner facing in one
> direction, there is a draught.

> If the chicks are evenly distributed around the
> brooder area, eating and drinking, the temperature is
> correct.

If you do use a thermometer, take the air temperature
50 mm above the litter.

Brooding is one of the most critical stages in growing chickens. The poultryman should make it a practice to
visit the brooder house often. This allows careful observation of what is going on and is also a stimulus to
the chicks to move about, feed and drink. Always visit
the chicks before visiting older birds. Never move between mature poultry and chicks without thoroughly cleaning
and disinfecting your boots and clothing.

If you find any deformed, sick or stunted birds, destroy them. They will probably never produce properly,
involve an inefficient use of your feed and facilities
and, if sick, may also present a hazard to the rest of the
birds.

Pasting

Occasionally the vent of a chick may become blocked by
faeces. This condition is known as pasting and generally

occurs as a result of stress such as temperatures which are too high or too low, poor nutrition, poor ventilation leading to draughts, diseases or too long a period without water between hatching and installation in the brooder. Management stresses such as dubbing, detoeing or debeaking may also lead to pasting. When pasting is observed, soften the pasted material with warm water and gently remove it. Ensure that the brooder conditions - temperature, ventilation, feed and water - are adequate and that the last two are readily available. A solution of 1% molasses in the drinking water or 5% molasses by weight in the feed for one day may help overcome the condition.

Impacted crop

Sometimes chicks will eat litter or other large, relatively non-digestible materials which will block the passage of food through the crop. This condition is known as an impacted crop or crop-bound. The crop may then become distended, either due to gas or because the chick continues to eat. Sometimes the impacted crop can be emptied by holding the chick upside down and massaging the mass and working the contents out through the mouth. Injecting small amounts of warm water into the crop by way of the mouth may also soften the mass so it can pass through. If these methods are unsuccessful, it may be necessary to cut surgically into the crop to remove the material. This generally is not worth the effort or risk unless the chick is from valuable breeding stock.

Do not allow visitors into your brooding or rearing operation. Visitors may bring disease organisms into the flock and the presence of strangers may upset the chicks. By the same token you should not visit other poultry operations nor exchange equipment with other farmers.

Changes in feeders and waterers in the house should be made gradually since chicks are creatures of habit and sudden changes may affect their feed intake and growth.

8.4 MANAGEMENT OF PULLETS

The word pullet refers to a female chicken up to the completion of its first laying season. Once it starts to lay at 20 to 21 weeks (the point-of-lay), however, its management is as described in Section 8.5 for layers. Only the period from the time the pullet leaves the brooder house at four to five weeks until it reaches the point-of-lay is discussed in this section.

Pullets can be reared on range, deep litter or in a battery which is composed of cages holding 1, 2 or 3 birds (see 8.5 for a description of the battery system). The battery system reduces the problems of disease and cannibalism and obviates the necessity for litter management. The initial cost is, however, much higher than for the deep litter system. In addition, the cages must be properly used and maintained in order to be successful. Pullets which have been reared in battery brooders will have little resistance to parasitic organisms such as coccidia so a high mortality may be encountered if they are later placed on litter. This is also true if battery-reared pullets are placed on litter when they start to lay. For this reason, chicks and pullets which are to be placed on litter as layers should also be reared on litter.

As with the chicks, the rearing house and all equipment should be thoroughly cleaned and disinfected before the arrival of the birds. In a deep litter house, clean, dry litter should be spread 100 to 125 mm deep over the floor and rounded into heaps in the corners. These heaps are necessary to stop the pullets from piling into the corners at night which may result in death of the birds due to crushing or suffocation. Pieces of plywood cut into equilateral triangles about 0.3 m on a side may also be leaned in the corners to prevent piling. Once the pullets are on the litter, the management of the litter is similar to that described for chicks (see 8.3).

Feeding equipment

For a deep litter system, one can use tube feeders or troughs, and for batteries troughs are used. These are shown in Fig. 45. Note that the side of the trough away from the bird is higher than that toward the bird to reduce wastage from feed being pushed over the edge. Having a lip curving in from each side will also reduce wastage.

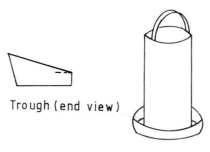

Trough (end view)

Tube feeder

Figure 45. Tube feeder and troughs for chickens.

Tube feeders should be suspended from the roof and raised as the birds grow larger. The outer edge of the feeder should be at or slightly above the tail height of the average bird in the house.

The pullets should be given fresh feed each day. When they are brought into the rearing house, the feed should be gradually changed from chick starter to growing mash. The pullets should be fed growing mash until they are about 17 weeks old when a gradual change to laying mash should begin. Feed troughs should not be more than one-third filled. This reduces wastage as well as ensuring that the birds always have fresh feed. Do not, however, allow feeders to remain empty for long periods or cannibalism may result. The birds will be encouraged to eat if the feed is stirred two or three times each day by running a stick or your hand through it.

Watering equipment

In a deep litter system, large fountain-type waterers

should be used. These are similar in design to tube feeders except that they are for water. Automatic waterers
are normally used in a battery house. When using automatic waterers utilising troughs, it is important that the
waterers be levelled with a spirit level so that all birds
have water at all times.

Roosts

Pullets which will be given roosts as layers should have
roosts in the growing house. The roosts can be built
along the side of the house, or in the middle in a system
known as 'half-and-half'. The placement of the roosts in
the two systems is shown in Fig. 46.

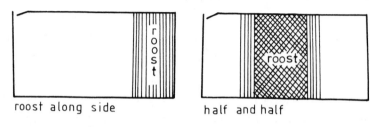

roost along side   half and half

Figure 46.   Roost placement.

As can be seen, in the half-and-half system, one half of
the floor space is taken up by the roost. The birds must
cross the roost to move across the house, thus encouraging
the use of the roost. Since some poultrymen consider it
desirable to have layers roost to prevent broodiness (see
8.5), it is useful to have the birds learn to roost as
pullets. The advantage of having the roost along the side
of the building is that cleaning is simplified if access
doors to the outside are provided. In either case, the
roosts will reduce soiling of the litter since most of the
droppings will be deposited while the birds are on the
roost.

The tops of the roosts should be about 760 mm from the
floor and framed of wood with the sides covered with wood
or metal sheeting. The top can be constructed of wood
slats 38-50 mm thick and 300 mm apart. If the roost is

along the wall leave a space of 300 mm between the wall
and the first slat. The slats should be made of cut lum-
ber with the corners rounded off. Round poles should not
be used as this puts undue pressure on the breastbone leading to crooked breasts or breast blisters. Wire mesh can
be used for part of the top of the roost. The most desirable system is to have wire mesh in the centre and slats
along the sides. If a dropping pit is used, wire mesh
should be installed under the slats to keep the birds out
of the pit. Whether the roost is along the wall or used
in the half-and-half system, it is desirable to have the
tops flat. This allows hanging feeders over the part of
the roost covered with wire mesh. Thus the floor (or
feeding) area is not reduced as it would be if the roosts
were slanted so that all the feeders had to be hung over
the floor. This is especially important in the half-and-
half system where half the floor space is taken up by the
roost. With slanted roosts, the birds will try to get as
high as possible for roosting so the upper parts will be
crowded and the lower parts under-utilised. With a flat
roost the birds will distribute themselves more evenly.
Each bird should be allowed 175 to 300 mm of linear space
on the roost depending on the size of the birds. A flat
roost is shown in Figure 47.

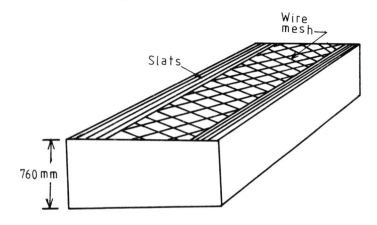

Figure 47. Design of a flat roost.

Range rearing

If pullets are to be reared on range they should be provided with shelters into which they can be locked at night or during inclement weather and which they can use to get out of the sun. The range used should not have had poultry on it for at least two years previously. Cultivation and planting a crop such as lucerne, clover or a grass mixture will help clear the area of disease organisms as well as providing feed for the birds. Stocking density on range should not exceed 1000 birds per hectare.

Light

As a general rule, increasing daylength stimulates production in chickens and decreasing daylength inhibits it. Since pullets which come into lay before they are mature will lay small eggs and will never produce as well as birds which start laying when they are mature, it is important to prevent them from starting laying too soon. In essence, this means ensuring that the birds are not exposed to increasing daylengths from about the eighth to twentieth week. At some times of the year, this is no problem as the daylengths are naturally getting shorter. In others, however, it may be necessary to start the birds on a long daylength and gradually decrease it until about the twentieth week. The rate of decrease should be about 15 minutes per week so the poultryman should determine what the daylength will be when the birds are twenty weeks old and then at eight weeks provide the chicks with this daylength plus an additional three hours. If the daylength is then shortened by 15 minutes each week, the daylength the birds receive in the 19th week should be equal to the natural daylength. If the days are still getting longer, the layers will then be stimulated to begin laying.

In general, the management practices used for pullets are similar to those for other poultry enterprises and, in particular, those relating to chicks (see 8.3).

Because there is no heating in the rearing house, it is important to take steps to keep the house warm on cold days.

As with chicks, any abnormal or sick birds should be isolated or culled. Poorly developed feathers are a sign of a poor growing bird or of litter being too dry. If the litter is correct, any birds which have under-developed feathers should be culled as they are an invitation to cannibalism.

Since growing pullets do not have the intensive labour requirement of brooding chicks nor the necessity for daily egg collection of layers, there may be a tendency to neglect the birds somewhat at this age. This is a serious mistake as the performance obtained during the growing period will directly reflect on performance during the laying period. It is also bad economics to go to the effort and expense of obtaining and brooding the chicks and then let them under-perform or die during the rearing period.

8.5 MANAGEMENT OF LAYERS

Pullets will begin laying small eggs at 20-21 weeks of age. These small eggs are known as pullet eggs and will reach normal size within one or two weeks. Once the pullets begin to lay, their management is the same as for hens in their second or later laying seasons.

Litter management

Layers can be housed on litter, in batteries or on range. In the litter system, the hens are run on 100 to 150 mm of litter on the floor of the house. Litter should be added until a final depth of 200 to 250 mm is reached. Management of the litter is the same as described for chicks (see 8.3). Care should be taken to see that the litter does not get too dry or too wet. The ideal moisture content is 20 to 30%. This can be tested by picking up a handful of the litter and squeezing it. If the

moisture content is correct, cracks will form in the
material when the pressure is released. If it forms into
a ball it is too wet and if it crumbles, it is too dry.
Litter that is too wet may lead to problems with coccidiosis or high ammonia levels in the house. If the litter
is too dry it may become dusty leading to respiratory
problems.

Nest boxes

It will be necessary to provide nest boxes in which the
birds will lay their eggs. Use of the boxes results in
cleaner eggs and less problem with breakage and egg picking by the birds. Each box should be 255 x 305 mm to 305
x 355 mm depending on the size of the hens. The boxes
should have good ventilation but should be kept dark. A
sloping roof over the nest boxes will keep the birds from
roosting on them. The nests should be placed about 0.5 m
from the floor with a perch at the front so that the bird
can light on it when it jumps up to the nest. It will be
necessary to provide a door at the front of the nest so it
can be closed at night to prevent the birds from sleeping
inside which is predisposing to broodiness. One nest
should be provided for every four or five pullets.

Pullets have to learn to use the nests. These should
be available to the birds as soon as they start to lay.
If the nests are placed crosswise to the long dimension of
the house the pullets will use them more easily. Installation of one or several china eggs in the nests will also
be an inducement to use them. Laying eggs on the floor
can become a habit so pullets which start doing this
should be induced to use the nest boxes. Eggs on the
floor should be collected immediately and the floor nest
filled in with coarse material such as whole maize cobs.
If pullets persist in using the floor, putting a nest box
near where they lay may get them to lay in the box. The
box can then be gradually moved toward the banks of nests
to induce the pullets to lay in the banks.

Nest boxes should be lined with material such as hay, straw, wood shavings or sawdust to reduce breakage of the eggs. If eggs become cracked or broken in the nest, this may lead to egg-eating by the birds: a very difficult habit to break. The normal procedure is to cull egg-eaters. Other possible causes of egg-eating include too much light in the nests, insufficient food and water or a lack of calcium in the diet.

Roosts

Roosts may be provided for layers in the same configuration as described for pullets (see 8.4). Opinion is divided regarding the need for roosts and many successful poultry operations do not use them.

Battery management

In the battery laying system each bird or pair of birds (or more) is placed in a cage. The dimensions of the cage will vary with the number and size of the birds housed. There are many configurations possible with the battery system ranging from a single tier of cages to triple tiers. A typical single tier house is shown in Fig. 48. Note that the cages are placed back to back with an aisle between the rows of cages.

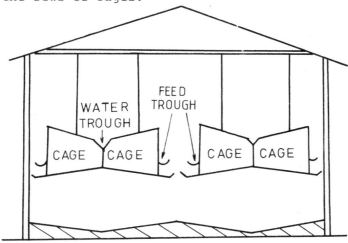

Figure 48. Typical single tier house.

The floor under the cages is sloped for drainage and easy cleaning. This, of course, is not possible unless a concrete floor is used. Provision should be made for protecting the birds from rain and extremes of heat and cold. A convenient method is to use hessian (burlap) curtains which can be rolled up or down depending on the weather.

A detailed drawing of a single back-to-back cage unit is shown in Fig. 49. Note that the wire floor of the cage is slanted so that the eggs will roll to the front. This serves to prevent egg pecking and to keep the eggs clean. Having the side of the feed trough away from the layer higher than the side toward it reduces the wastage due to feed being pushed over the outer edge. Inward curving lips should be provided on both sides of the trough.

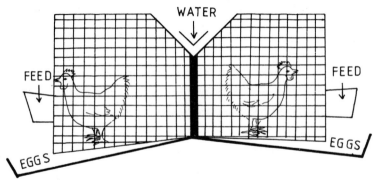

Figure 49. Back-to-back battery cage unit.

Although the initial cost of the battery system is much higher than the deep litter system, it reduces the incidence of cannibalism and egg pecking, reduces contamination of food and water with droppings, allows easy recording of individual egg production, reduces bullying or interference at the feed or water troughs, reduces the spread of disease and results in clean manure which can be dried and has been used as a non-protein nitrogen source for feeding ruminants. The deep litter system is cheaper and easier to establish and operate than the battery system; there usually are less fly problems and egg production is somewhat higher.

Free range management

Free range management is generally not used in commercial operations. It requires much more labour than the more intensive systems as well as a substantial investment in fencing. Predators can be a problem and eggs not laid in the nests provided may never be found. For the small operation however, a small area where the hens can be pastured may be suitable as long as adequate shelter, nests, food and water are provided.

Feeding layers

All layers should be fed laying mash or pellets. Hens will eat more pellets, and therefore will lay more eggs (up to a point) if fed pellets. Pellets, however, are more expensive than mash so the benefit in terms of profit is generally marginal. Fresh concentrate feed and fresh, clean water should be given to the birds each day. In addition, provision of fresh green feed (3 kg/100 birds/ day) and oyster shell will improve egg production and quality. Be sure that green feed offered to the layers is not contaminated by germs, parasite eggs or larvae or by chemicals. One tube feeder or a single 2 m long trough containing oyster shell (calcium carbonate) should be provided for each 100 birds.

Birds do not have teeth. Their feed is ground between hard materials contained in the muscular gizzard. Where very coarse feeds such as whole or cracked grain are being fed or when birds have access to litter or feathers it may be necessary to provide insoluble grit (usually flint or granite cracked to a specific size). Grit should not be fed *ad lib.* as the birds tend to consume much more than they need. This passes through the digestive tract without harming the bird but is a waste of money. Provide 500 gm for each 100 birds each week in their feed troughs.

The birds should never be allowed to be without feed or water as this may cause them to moult. Moult may also

be caused if the ventilation of the house is inadequate.
There must be plenty of fresh air in the house although no
birds should be exposed to direct draughts.

Light

The length of daylight is an important factor in egg production. As a general rule, increasing daylength has a
stimulatory effect on production whereas decreasing daylength will inhibit production. If the hens come into lay
during the longest daylength of the year, artificial means
will be required to maintain this daylength whereas if
they come into lay during the period of shortest daylength
no artificial light will be required. If desired, the
daylength may be maintained at a constant 14 to 16 hours
of light and will give good results. Extra daylength can
be provided using artificial light from electric bulbs or
lanterns. The effect of the extra daylength is to provide
extra hours for eating, etc. and also seems to result from
a stimulatory effect on the pituitary gland. If long daylight hours are to be maintained, it is necessary merely
to turn on the lights at dusk and leave them on until the
desired hours of light have elapsed. Increase the length
of daylight gradually - not more than 15 minutes per week.
If using electric lighting, it is most convenient to use
time clocks to turn the lights on and off. The intensity
of the lighting does not have to be anywhere near that of
daylight to get an effect.

Broodiness

A broody bird is one that suddenly stops laying and begins
to set her eggs. Broodiness is generally brought on by too
much warmth in a deep litter house, overcrowding or by the
birds sleeping in the nests. The condition can often be
overcome by isolating the broody fowl in a shaded place
for several days where she is exposed to plenty of cool,
fresh air. If possible she should be placed in a small
cage with a wire or slatted floor. The broody bird can be

recognised because she walks with her wings slightly raised and her head down. After the broodiness disappears and the hen begins to lay (within two or three days), she should be returned to the flock. The heritability of broodiness is very high, so birds which go broody should not be used as parent stock for future generations.

Flying hens

Hens which cause problems by flying either within the house or out of outside runs can be stopped by clipping the flight feathers of one wing. This will unbalance the bird so she is unable to fly. To clip the wings, cut off about half of each of the major flight feathers with a pair of scissors.

Culling layers

Laying mash is expensive and a bird which is not laying or is sick is not paying for the feed she is eating and so should be culled. Identification of non-productive birds is easy in a battery system if each cage is numbered and a record is kept of the eggs produced by each bird each day on a form such as that shown in Fig. 50. It is then easy to decide which birds to cull. It is generally considered that a layer producing at less than 50% (i.e. less than one egg every second day) is not earning her keep. In a deep litter system, the level of productivity must be determined by trap-nesting, i.e. catching the birds in the nests when they enter to lay their eggs, or by examining the birds for the various signs of lay. The signs of lay are as follows:

> A non-laying bird has a yellow pigmented skin if on a diet with high carotene levels. As lay progresses, the yellow pigment of the skin is lost. The loss is first apparent around the vent, then the eye rings, followed by the ear lobes and beak and finally the shanks. It is believed that the yellow pigments go into the production of the yolks of the eggs.

221

DAILY EGG PRODUCTION RECORD

Month . . . . 19 . . . .

House Number . . . .

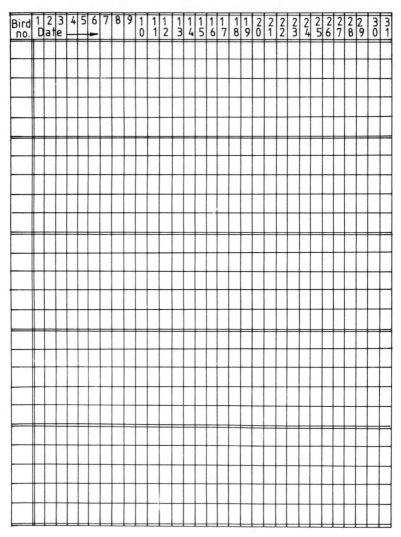

Figure 50.   Daily egg production record.

A hen in lay has large, prominent, awake, bright eyes.

The comb of a laying bird is large and is a bright, glossy red. Both the comb and the wattles are warm and feel waxy.

The vent is large, oblong, moist and faded.

The pelvic bones are thin, pliable and separated. It
should be possible to place three or four fingers
between the bones.

The plumage of a good layer will appear ragged and
dirty.

The proper management of layers follows the same rules as
laid down in the sections for chicks and pullets. Remember that your layers are highly strung, nervous animals
but they can produce money for you, so you must handle
them with care.

8.6   MANAGEMENT OF BREEDING STOCK

At times it may be desirable or necessary to breed and
hatch your own chicks. The decision to do this should not
be made in the middle of the laying season. If you plan
to rear your own chicks this decision should be taken when
the parent stock are one day old or even before they are
delivered. Such an early decision is necessary because
the brooding and rearing management of breeding stock is
somewhat different from that for production birds.

Pre-breeding management

When brooding breeding stock, at least 15 cockerels should
be included for every 100 pullets. This number can be reduced to 12 at eight weeks when any obviously poor cocks
should be removed. A further culling at 22 weeks, just
before breeding begins, can reduce the number to 8 or 10
for the heavy meat breeds. For light breeds such as the
Leghorns, 15 cocks per hen should be used and for the medium breeds, such as the Plymouth Rock, about 12 is sufficient.

Dubbing and detoeing.   At 1 day, both the males and fe-
males should be dubbed (see 8.10) and the inside and back
toes of the males should be removed at the outer joint.

This can be done with a debeaker or with a special detoer which consists of two hot wires which come together and cut off and cauterise the toe. Detoeing serves to prevent injury to the hens during mating as, when the male mounts the female, he pushes on her back with his feet (strides). If the cock has long, sharp toenails he may injure the female.

Females should also be debeaked when one day old (see 8.10).

At 12 to 14 weeks the cocks should be dewattled. This should not be done later as the stress, especially if done just before breeding, may interfere with fertility. If detoeing was not done during the first week, the toenails can be blunted just before the cocks are placed with the hens. They should not be detoed at this time, however.

Installation in the breeding house. Establishment of a social order among the cocks is very important and the poultryman must ensure that this does not interfere with breeding. If possible the cocks should be installed in the breeding house for a few days before the pullets arrive. This will allow the males to become used to each other, to establish a social ranking and to get used to the house. If it can possibly be avoided, cocks should not be introduced after this initial installation as this will lead to fighting among the cocks which may interrupt breeding and may also result in death or injury. If it is necessary to add cocks, this should be done about one hour before dark so the birds have a short period before roosting to become habituated. By the time the roosting period ends the next morning, the new male should be established in the social order with minimal stress.

Management during breeding

Flock mating. A cock will mate 20 to 80 times each day depending on competition, the number of females available, the cock's position in the social order and climatic factors such as temperature and light. Most of these matings will occur early in the day. Cocks may mate several times with a certain hen and it has been shown that even in a flock, individual males mate only with certain females. If for some reason a cock is unable to mate, the females which he has been mating will not accept another male until the first cock is removed. Having too many cocks in the breeding pen will reduce fertility since there will be undue fighting and too much competition for the available females.

Pen mating.   Pen mating with only one male per group of hens can also be carried out. This has the advantage that the parentage of the resulting chicks is accurately known but does represent an increased capital investment for the breeding pens. Pen mating has the obvious disadvantage that, if a male is sterile, none of the hens to which he is mated will produce fertile eggs. It has also been shown that the overall fertility when using pen mating is lower than with flock mating even when fertile males are used in both situations.

Selecting breeding stock

When selecting breeding stock, only those birds with good vigour should be used. This means that they are active, alert, have interest in the opposite sex, eat well and have full crops. Vigorous birds also tend to go to the roost late and get off early in the morning. In a study where cocks were judged to have low, average or good vigour, there was a doubling in the number of matings per day between each classification. Thus, the cocks judged to have good vigour performed four times as many matings each day as the cocks with low vigour. Breeding stock should have good plumage and good body conformation for the improvement of the chicks. As previously mentioned, selection of breeding stock should take place during the growing period as well as just before the birds are placed

in the breeding pen. Good selection of breeding stock
will improve the quality of the chicks produced.

Selection of eggs for incubation

The hatchability of the eggs is greatly affected by conditions in the breeding pen. Vigorous, healthy pullets
will produce eggs of higher hatchability than hens. The
eggs should be laid in clean nests and be free from cracks
and dirt. Eggs to be incubated should not be cleaned.
The eggs should be between 55 and 60 gm. Eggs which are
lighter or heavier than this will result in a lowered
hatch. The eggs should be of normal shape with shells of
medium strength. If the shell is too thin evaporation
from within will be too rapid, whereas if it is too thick
the chicks will have difficulty pipping the shell at
hatching.

Refer to Section 8.9 for a discussion of storage and
handling of eggs for incubation and of the procedures for
incubation.

8.7 MANAGEMENT OF BROILERS

Broilers are birds which are bred and raised especially
for meat. Management practices up to 5 weeks are as described in Section 8.3. After the brooding period, the
broilers are normally raised on litter using the 'batch'
system. Under this method, a house will be filled with
the proper number of birds (allowing 0.1 square metre of
floor space per bird) which are then reared together and
all killed or sold at nearly the same time. This is normally when the birds weigh 1.4 to 2.0 kg liveweight and
are between 9 and 11 weeks old.

Housing, litter management and feeding and watering
equipment requirements are similar to those for pullets
(see 8.4). The tube feeders should be suspended from the
rafters so that the feed pan is level with or just above
the tail of an average bird in the house. As soon as the
broilers reach five to seven weeks of age they should be

started on broiler mash. This is a feed which is formulated to promote rapid growth. The feed should always be fresh and no accumulation of feed should be allowed in the bottom of the feed pan.

The general management and handling procedures for broilers are the same as for pullets (8.4) and layers (8.5) except that broilers should not be caught with a catching hook since their greater weight may result in broken legs.

It is possible to improve the growth rate and fattening characteristics of cockerels by surgically castrating them or implanting female hormones under the skin but some countries forbid these practices. Cockerels which are thus treated are known as capons.

Marketing of broilers
_____

The most convenient means of marketing broilers is to sell an entire batch either live or dressed. For shipping, live birds should be packed into crates in which they will be protected from stresses such as bruising, over-crowding, or temperature extremes. If the birds are to be sold dressed, you should contact the buyer to determine the method he prefers. The following describes one method for preparation of birds for the table.

Dressing poultry for table.   Starve the birds 12 to 24 hours before killing. The birds can be killed by chopping off the head, dislocating the neck or cutting the jugular vein through the mouth. Cutting off the head with an axe is easy but is not recommended as the skin of the neck will contract so part of the neck is exposed. This will dry out and be unusable. The neck is dislocated by pulling the head straight out and snapping it back. A small slit in the skin of the neck will allow the blood to drain. The quickest and easiest method (once the technique is learned) is to insert a knife into the mouth to cut the jugular vein and to pierce the hind-brain. The bird should then be hung upside down to drain. This method provides for good blood drainage and is hygienic but you should not attempt it unless the technique has been demonstrated for you. Piercing the hind-brain also relaxes the feather follicles so the bird can be dry plucked.

Whichever method is used, the bird should not be allowed to flutter about as this will result in bruising of the carcass. This can be prevented if the bird is killed in a bucket or receptacle or is hung by the legs after killing.

Birds can be plucked dry or after scalding. Dry plucking can only be used when the hind-brain has been pierced as this leads to the relaxation of the feather follicles. The bird should be plucked immediately after killing, starting with the wing and tail feathers then proceeding to the breast and neck. The feathers should be grasped a few at a time and pulled in the direction of growth. If too many are pulled at once, you may tear the skin. You can scrape off the pin feathers with a blunt knife.

Scalding also serves to relax the feather follicles before plucking. The carcass should be soaked in cold water, drained, and then dipped in hot water (53-70°C) until the feathers loosen (30-60 sec). Broilers will require the lower temperatures for a shorter period while older birds require higher temperatures for a longer period. Scalding should be done immediately after bleeding so the feathers do not 'set'. The order of plucking is the same as for dry plucking.

Some commercial poultry dressing plants dip the carcasses in a molten solution of 68% paraffin wax, 30% resin and 2% fat. The carcasses are then sprayed with cold water to harden the coating which is then peeled off, taking the pin feathers with it. This is done to get rid of the pin feathers and fine hairs after the majority of feathers have been removed.

After the feathers and pin feathers have been removed,

there will still be many fine hairs on the carcass. These can be removed by passing the carcass quickly through the flame of methylated spirits or a similar compound. Paper or candles should not be used as they will leave a carbon deposit on the carcass.

The sinews can be removed from the leg by breaking the shank below the hock and twisting it. If the skin does not completely break off, it should be cut with a knife (don't cut the sinew!). If you then hold the thigh and pull on the shank, you will pull the sinews out. Removal of the sinews is unnecessary for young broilers as the sinews are tender and do not detract from carcass quality.

To remove the internal organs of the bird, cut from the base of the neck to where the head was removed. Cut off the neck against the back (do not cut the skin) and remove the crop, oesophagus and windpipe. Hold the carcass in a sitting position and loosen the heart, trachea and lungs. Slit the abdominal region from the point of the breast bone to, and around, the cloaca. Use your finger to loosen the membranes holding the intestines. If you then firmly grip the gizzard and pull, you can remove all the internal organs except the sex organs which have to be separately detached from the back.

The liver, gizzard and heart comprise the giblets which can be cooked with the carcass or separately. Carefully detach the gall bladder from the liver and discard it. Slit the gizzard, remove the contents, and rinse thoroughly. The heart should be carefully trimmed. The cleaned giblets should be replaced with the neck in the abdominal cavity.

The wings can be twisted behind the back to provide a firm base for the carcass in the roasting pan, and the body cavities stuffed. The neck cavity can be closed by pulling the flap of neck skin up and pinning it to the breast. The abdominal cavity can be sewn shut.

You can expect 9-12% of the liveweight of a bird to be blood and feathers, 16-23% to be the head, feet and viscera. The drawn weight, including the giblets, should be 65-75% of the liveweight. Of this, 49-60% will be edible meat and the rest, bone. The dressing percentage increases as the liveweight increases.

8.8   HANDLING FRESH EGGS

The eggs produced by your layers are the primary product of your laying flock. The price you receive for your eggs will depend to a great extent on the quality of the eggs produced and marketed. Quality of eggs depends on their freshness, cleanliness and shell condition. These are, in turn, influenced by the conditions in the laying house when the eggs are laid and how the eggs are collected, stored and marketed.

Having too few nest boxes or boxes which are too small will lead to cracked eggs due to eggs knocking against each other. Cracking will also occur if the eggs are not collected at frequent intervals. Collections should occur three or four times each day. Morning collections are especially important since 70% of the hens lay their eggs within the first five hours after bright light begins and 90% within seven hours. Thus, if daylight occurs at 5 a.m., 90% of the eggs will be laid by noon. The nest boxes should be cleaned regularly and provided with clean litter to act as an absorbent and to provide a cushion for the eggs. Broody hens should be kept away from the nest boxes as they will move the eggs about and may thus crack them. One should also watch for the hen which pecks the eggs as this also results in cracks and deterioration. Male chickens must be kept away from the laying flock because fertilised eggs will not keep as long as infertile eggs and may show blood spots (the developing embryo).

Eggs should be collected into wire baskets or egg trays. They should be handled carefully and placed, not dropped, into position. If placed in egg trays, the big end should face upwards. Egg baskets should not be filled more than half full. After collection, the eggs should be

taken to a cool area and cleaned with steel wool, sand-
paper or a damp cloth. Wahing eggs in water is not advised as the water and dirt may penetrate the shell. The
good quality and appearance of the eggs will be retained
longer if the eggs are lightly sprayed with a neutral oil
such as vegetable or olive oil after they are cleaned.

Top quality eggs should weigh at least 58 gm, should
be of uniform size and not have cracked shells. Small
eggs may be the result of pullets laying small eggs,
poorly developed pullets, deficiencies in the diet, or
temperatures which are too high or too low. Once the eggs
have been graded for quality they should be packed on
trays for storage and marketing. Store them in an area
with a temperature of 10-15°C with a relative humidity of
about 75%. Keep the eggs away from paraffin (kerosene),
onions, fish, potatoes and other products from which the
eggs may absorb odours.

Eggs may be marketed by on-farm sales or deliveries to
individuals, sales to shops and stores or sold to whole-
salers. However you market your eggs, the profitability
of your enterprise depends on getting and maintaining a
reputation for producing high quality eggs. This is ensured if you collect and handle your eggs in the proper
manner and market them as soon as possible after they are
laid.

8.9   INCUBATING EGGS

Since it is possible successfully to transport day-old
chicks long distances by road, rail, or air, it is the
common practice for the broiler or egg producer to buy his
stock from firms which specialise in the incubation of
eggs. At times, however, it may be desirable to hatch
chicks on the farm. The easiest way to do this is to provide a broody hen with a clutch of 10 or 12 fertilised
eggs and allow her to carry them through to hatch. If a
broody hen is not available or a large number of chicks is
required, artificial incubation may be used.

There is a wide variety of small incubators available which may be heated by gas, electricity or paraffin and have either still air or forced draught ventilation. The primary purpose of an incubator is to provide the right conditions of temperature, humidity and air movement to maximise the proportion of the eggs which will hatch. The forced air incubators are usually designed so that most operations (including turning the eggs) are automatic, whereas turning the eggs, ventilating, etc. must be done manually with the still air machines. Whichever you use, you should follow the manufacturer's instructions carefully and whenever the incubator is in use, keep the instruction book nearby.

The following steps should be carefully followed when incubating eggs in a still air incubator. (The procedure is the same for forced air except for those steps which are done automatically by the machine.)

Preparation of equipment

The incubator should be located in a room which can be maintained fairly constantly at 22°C and where the incubator will not be exposed to direct draughts. The room should, however, be well ventilated.

The room, incubator and any other equipment should be thoroughly cleaned and disinfected. Several days before installation of the eggs, start the incubator to be sure it is operating satisfactorily and to allow all the parts of the incubator to come to thermal equilibrium. During the first week of incubation the temperature should be 37.5°C, during the second and third 38°C. Reducing the temperature slightly during the last two days of incubation (days 20 and 21) may increase hatchability. The required temperature is closely related to the humidity. Ideally, the relative humidity should be about 60% at the temperatures quoted. Most manufacturers have determined the ideal temperatures to be used in their incubators and the instructions provided should be carefully followed.

Installation of eggs

The eggs to be incubated should be carefully selected from
hens which are of good genetic stock. The eggs should be
normally shaped and weigh 55 to 60 grams. The shells
should be smooth with a good bloom but not shiny and be
without cracks. You can check for invisible cracks by
gently knocking the eggs together. A cracked egg will
give a dull non-metallic sound. Cracks can also be detected by candling. Eggs for incubating should be as
clean as possible but should not be washed. It is therefore necessary to provide clean nest boxes and frequent
collections when producing eggs for incubation. The eggs
should not be stored for more than seven days before being
installed in the incubator as the older eggs do not hatch
as well as fresh eggs. The eggs should be stored at 13°C
and turned daily. Turning involves rotating back and
forth through 90°. This can be done conveniently if the
eggs are placed in an egg tray. A 25 mm block can then be
placed under one end. The next day it should be removed
and placed under the other end. At higher temperatures,
evaporation is too rapid, and at lower temperatures, the
embryos may be killed. The eggs should be stored and
placed in the incubator upright or slanting with the big
end up. If the small end is up, hatchability will be reduced.

Incubating procedure

The eggs should be turned regularly (at least 3 times per
day) from the third to the nineteenth days of incubation.
Turning serves to prevent the developing embryo from
sticking to the shell. After the seventeenth day of incubation the chick starts pipping (cracking the shell) and
moving itself. When turning, the eggs should also be
moved about in the incubator to obviate the danger of the
same eggs being continually exposed to cold or hot areas
in the incubator. Turning should involve only a movement

from 45° from vertical on one side to 45° from vertical on
the other. Turning should be done up to six times per
day. It is of definite value in still air incubators to
allow the eggs to cool while exposed to fresh air each
day. As the eggs cool, the fresh air is drawn through the
shell thus facilitating respiration by the embryo.

Candling eggs.   The eggs should be tested (candled)
twice during the incubation period. For candling, the
eggs should be held in front of an intense light source in
a darkened room. This allows the operator to look through
the eggs to study development.

At seven days, the normally developing chick will
resemble a spider with a dark spot (the embryo) in the
centre resembling the spider body with the veins radiating
from it resembling the legs. An infertile egg will appear
clear with no embryonic development. Dead embryos will be
encircled by a dark ring or will appear only as a black
spot without veins. This spot generally will not move
when the egg is rotated. At 14 to 18 days, the live chick
will appear as a dark mass with veins apparent near the
air space. The dead embryo has no blood vessels and will
move freely when the egg is rotated.

Effects of temperature.   If the incubator is too hot the
embryos will develop too early, the proportion of abnormal
and dead chicks will be increased, many chicks will pip
but be unable to break the shells, the chicks will be
small and the navel will not be properly healed. If the
incubator is too cold, hatching will be delayed resulting
in a low hatch with many abnormal embryos. The chicks
will be large but have a low viability.

Effects of humidity.   If the humidity is too high, the
growth of the embryo is accelerated with a high mortality
on the nineteenth day. This is probably due to a change
in the composition of the yolk sac and a restriction of
the movement of the embryo and its respiration. If the

humidity is too low the development of the embryo will be
delayed and the chicks will be small with short down.

Movement to hatching trays

At 19 days the eggs should be moved to trays where the
chicks will hatch. The chicks should be left in the
hatching compartment until they are dry. They should then
be removed and placed in a brooder house (see 8.3). If
high quality eggs and the proper procedure are used, at
least 60% of the eggs should hatch in a still air incubator and 70% in a forced-draught incubator.

Incubation is an art and with experience, as you learn
the characteristics of your machine, your hatching percentage should increase.

8.10 CANNIBALISM

Cannibalism means eating a member of the same species. In
most of the mammalian domestic stock cannibalism is confined to eating of the young by the mother. In chickens,
however, cannibalism among stock of all ages may be found.
It can occur in brooding chicks, growers or layers. As
with egg pecking or eating, this vice is contagious and
when one bird starts, others are liable to pick up the
habit. One cause of cannibalism is prolapse of the oviduct after a pullet lays an egg. Since the oviduct
appears red and moist, it will attract pecks from the
other birds and may result in death of the pullet. Other
causes of cannibalism may include overcrowding, over-
heating, poor ventilation, inadequate feeder or waterer
space, insufficient fibre in the feed, use of too much
yellow maize or the use of pelleted or crumbled diets.

Stopping cannibalism

There is a number of steps that can be taken to stop
cannibalism. Dimming the lights, darkening the windows
and using red bulbs may be of value in a growing house.

Any birds that have been pecked should be removed or the wounds painted with a tar or anti-pick salve. This will help with healing as well as making the wound unpalatable to the remainder of the birds.

Specs. Commercial devices known as specs or bits are available for birds which persist in cannibalism. The specs prevent forward vision by the bird so she cannot see what she is picking at. The bits cover the beak so the bird will be unable to see unless the head is in the eating position. The application of these varies but generally involves inserting a pin through the nostrils to hold the bits or specs in place.

Debeaking. The most common cure (and prevention) for cannibalism is debeaking, but there is little agreement as to the best age to do this. The aim of the process is to cut the beak so it will not grow back to its normal length while creating as little stress on the bird as possible. Many chicks are debeaked at one day of age but it is difficult to obtain uniform results as the beak is too small. This also may interfere with the chick learning how to eat. On the other hand, early debeaking causes less stress as a direct result of the operation. The beak of the day-old chick may be cut off with a guillotine blade with no heat added, or with a hot blade. The latter is preferable as it cauterises as it cuts and will prevent regrowth of the beak.

A more usual time for debeaking is at 6 to 9 days of age. This must be hot-type debeaking and, if done properly, will prevent regrowth of the beak. To debeak at this age the beak of the chick is inserted in a 44 mm diameter hole in the debeaker. A hot blade (at 815°C) drops down to cut the beak. The beak must be kept in contact with the blade for three seconds. Too short a time will lead to regrowth of the beak and too long a time will cause excess stress. Both the upper and the lower beaks are cut, but the head should be tilted so the upper beak is cut slightly shorter than the lower.

Pullets may be debeaked at 10 to 14 weeks or at 18 weeks. The procedure is similar to that used for 6 to 9 day debeaking. The beaks (both upper and lower) should be cut 45 to 65 mm in front of the nostrils. It is important that a cauterising cut be used and that the beak remains in contact with the blade for sufficient time to prevent regrowth. If the beak is quite hard, it may be necessary to cut one beak at a time.

Debeaking is stressful and should not be combined with other stresses such as vaccinations, high temperatures or changes of feed. When you debeak, supply an increased level of feed and water immediately after and be sure that the birds are able to eat and drink. This is especially important with nipple type drinkers as birds with sore beaks may be unwilling to push on the nipple to obtain water.

Dubbing.   In breeds such as the Leghorn which have large combs, these may become targets for pecking. To prevent this, the combs are usually removed at one day of age, a process known as dubbing. With day old chicks this simply involves cutting off the comb with a small pair of scissors. Both the comb and wattles are removed from cockerels, though the wattles are normally removed at about 12-14 weeks. In older birds, if the comb has been picked by other birds, or as happens sometimes, it has been frozen or has been torn in a battery cage, it may be necessary to remove the comb. This is a much more severe operation than dubbing at one day and care will have to be taken to reduce bleeding. The comb should be cut with dull shears that will crush as they cut. The wound should then be cauterised with a hot knife and treated with an astringent (a compound which stops bleeding). The bird should be isolated until the wound heals.

8.11   POULTRY DISEASE

This is a very broad topic and it is not the intention
here to list all possible diseases of poultry. The intent
is rather to describe some of the steps the poultryman can
take to prevent disease in his flock or to handle an outbreak should one occur. The causes of diseases to which
poultry are susceptible include dietary deficiencies,
parasites, bacteria, viruses and physical injuries such
as broken bones.

The most common dietary deficiencies are a lack of
calcium which results in poor quality egg shells and a
deficiency of vitamin D which leads to crooked legs in
chicks. Normally, commercially compounded rations will
provide adequate amounts of the vitamins. If additional
calcium is needed, it can be provided in the form of calcium carbonate (oyster shell) in feeders in the house. In
areas where the basis for poultry feed is white maize,
there may be a deficiency of the vitamin A precursor -
carotene - in the diet. This will result in eggs with
pale yolks. Provision of green plants in the laying house
will overcome this deficiency.

Parasites include roundworms, tapeworms, fleas, mites
and ticks. Roundworms are difficult to control on range
or in a deep litter system although in the latter case if
clean litter is used and only chicks free from roundworms
are brought in, there should be no build-up. Monthly use
of a piperazine-based antihelminthic in the drinking water
is an effective means of prevention and control. Fleas,
mites and ticks are best controlled by scrupulous cleanliness to prevent their introduction. If they do get into
your poultry units a variety of dusting powders are available for their control. The mites and ticks which infect
chickens are more difficult to control than the ticks
which infest mammals since the former spend a large portion of their life cycles off the host in the nests or in
cracks in the building. They have the ability to go long
periods without feeding so if the house is not thoroughly

cleaned between batches of birds, de-population in itself may not solve the problem.

The coccidia are a serious protozoal parasite of poultry. These spend part of their life cycle in the chicken and part in the litter. During the latter period they rely on fairly high moisture levels for continued development so it is very important that litter is not allowed to become wet. This is especially true around waterers. If the litter should become wet in these areas the wet litter should be removed and new, dry litter put in its place. Many commercial feeds contain coccidiostats at low levels which help to keep the coccidial populations under control. It is also possible with good management to promote resistance in your birds.

Common bacterial diseases of poultry include fowl cholera and fowl typhoid. Newcastle disease and leucosis are caused by viruses.

Control methods for poultry diseases include vaccination and isolation, treatment or destruction of infected birds. You should consult your veterinarian to establish a disease control programme for your flock. This will depend on the diseases which are endemic in your area and the vaccines available. The establishment of such a programme is well worth your time and effort as a single outbreak may put you out of business. If, for example, in some areas, Newcastle disease should break out in your flock, the entire flock will be destroyed and you will be enjoined from having poultry on the premises for a full year.

In addition to vaccination, the primary means of preventing disease in poultry are cleanliness and the prevention of introduction from external sources. The following are some of the steps to be regularly followed to maintain a high standard of cleanliness:

Regularly wash and disinfect all feed troughs and waterers.

Clean perches and roosts regularly.

Regularly stir deep litter to promote bacterial action to break down the manure.

When vaccinating, be sure all equipment is thoroughly sterilised.

Whenever a house is emptied of birds it should be cleaned, scrubbed and disinfected.

All woodwork should be creosoted and all walls white-washed annually.

Poultry houses should be enclosed with wire so wild birds cannot enter. This also results in a saving in feed.

Try to eliminate rodents (rats and mice) from your farm. These animals consume and waste an amazing amount of feed as well as carrying disease and parasites from house to house.

Prevention of the introduction of disease from external sources is best accomplished by refusing any outsider access to your birds. If outsiders must enter your premises, insist that they wear clean overalls and disinfect their boots before entering. Also, do not go from your mature birds to brooders without changing your clothes and thoroughly disinfecting your boots. Do not visit other poultry units and do not borrow or loan poultry equipment.

Symptoms of disease

There are a number of signs of sick birds. The following are some of the signs you may see:

Feathers look dull, broken or rough.

Tail feathers are dirty and matted with faeces or may appear curled.

Loose or bloody faeces are seen.

The bird may isolate itself or stand on one leg.

The bird may stand crookedly or appear to be straining.

The beak may be bloody.

The eyelids may be droopy or closed or the shoulders may be hunched.

There may be no appetite.

Whenever a sick bird is seen, it should be isolated from the other birds. You should then contact a veterinarian and follow his recommendations for treatment. One sick bird today often means many sick birds tomorrow.

Supportive therapy

While the veterinarian will advise the medication that should be used with sick birds, there are also steps you, as the poultryman can take to help sick birds recover and to prevent other birds from becoming infected:

Be sure that there are adequate supplies of clean, fresh water in the house. If necessary (or if possible) add more waterers so sick birds do not have to travel very far for water.

Add feeding places and feed small amounts often. This will help to keep the feed fresh and palatable and your regular attendance will encourage the birds to eat.

Feed a diet low in fat as fats are hard to digest. Addition of whey, molasses or extra maize will increase the energy content of the feeds.

Keep the air in the poultry unit fresh while avoiding draughts and chilling.

These steps are known as supportive therapy and basically represent the effort of the poultryman to help the chicken fight its battle against disease organisms or stress.

As a final word on poultry disease it must be reiterated that most diseases can be prevented by proper management. This includes the housing, feeding, watering,

vaccinations and other management steps which are taken
to ensure high flock productivity. If an outbreak does
occur, you must be able to spot it early and know what
steps to take to stop it. Whenever you are in doubt you
cannot afford not to contact a veterinarian. Many poultry
diseases are fatal and highly contagious and can completely ruin your poultry enterprise.

Appendix 1: Glossary of Livestock Terms

Abort: to expel the foetus prematurely.

Abscess: a localised accumulation of pus or matter in a cavity formed by the disintegration of tissue.

Afterbirth: membranes surrounding the foetus in the womb. Is expelled after the foetus during the birth process. Also known as the placenta.

Antiseptic: a substance used to kill harmful organisms on the skin surface.

Artificial insemination (A.I.): The technique of artificially introducing the semen of the male into the reproductive tract of the female.

Avian: a generic description of birds in general.

Bag-up: filling and distension of the udder prior to parturition.

Barrow: a male pig which was castrated while immature.

Billy: an intact, mature male goat.

Bloat: the build up of gas in the rumen of cattle, sheep, and goats. The gas is usually entrapped in bubbles and cannot be expelled normally by belching. Also known as hoven.

Bloom: the flush, fresh, satiny appearance of the coat of a healthy, well-fed animal.

Boar: an intact male pig.

Bovine: a generic name for cattle.

Brand: a marking placed on the hide of an animal by applying extreme heat, cold, paint or caustic materials.

243

Breed: a group of animals with distinct shapes and colours which produce offspring with similar shapes and colours.

Broiler: a chicken raised especially for meat purposes and marketed at about 1 kg dead weight.

Broken-mouth: a mouth having teeth missing. Usually applied to sheep or goats and occurs with old age.

Broody: refers to a hen in reproductive condition which attempts to incubate its eggs.

Browse: fodder obtained from eating leaves and twigs of bushes.

Buck: an intact mature, male goat or sheep.

Buckling: an intact, immature male goat.

Burdizzo: an instrument used for bloodless castration which clamps off the tissue connecting the testis to the rest of the body.

Bull: an intact male bovine.

Bulling: describes a cow on heat or the act of service by the male.

Calf: a young bovine of either sex.

Calve: to give birth to a calf.

Calving interval: the length of time from one calving to the next.

Camp: a pasture or grazing area.

Cannibalism: the act of eating another member of the same species. Seen in poultry and swine.

Capon: a male chicken whose reproductive organs have been removed or rendered inactive while the individual is still young.

Caprine: a generic name for goats.

Carcass: the dressed body of an animal or deal animal.

Cast: to place an animal in a prone position - usually for examination or treatment; also, to reject from the flock.

Castrate: to remove the testes of the male or to render them inactive; alternatively, an animal whose testes have been removed or rendered inactive.

Cattle: animals of the family Bovidae, genus Bos.

Check: any period of reduction of the normal growth rate. Usually caused by disease or poor management.

Cleanse: to expel the afterbirth and other waste from the female reproductive tract after giving birth.

Clip: to cut hair from animals; also, the total of the wool shorn from a flock.

Clutch: the total eggs produced by a hen on consecutive days.

Cock: mature male chicken.

Cod: the scrotum (in the castrate, the cod is usually filled with fat).

Colt: a young, male horse.

Cockerel: young, growing male chicken.

Concentrate: feedstuff low in fibre and high in digestible nutrients.

Corral: a small, fenced yard for confining livestock (khola, kraal).

Cow: a mature, female bovine having had at least one calf. (N.B. a cow between her first and second calves is often known as a 'first-calf heifer'.)

Cream: the portion of milk which has a high fat content.

Creep: an enclosure to which only the young of the species have access so they may be fed separately from the adult stock.

Creep feed: to provide special feed for the young; also, the feed provided for the young within a special enclosure.

Crossbred: the offspring resulting from the mating of a male and female of different breeds.

Crush: a long set of parallel panels spaced just wide enough for one animal to walk through at a time. Used for vaccination, examination, treatment, enumeration, sorting, etc.

Crutch: to remove soiled wool from between the hind legs.

Cud: roughage regurgitated from the rumen for chewing.

Cull: to dispose of the poorer animals in a herd or flock.

Dagging: removal of soiled wool from a sheep's hind-quarters.

Dam: the mother of an animal.

Deadweight: the weight of an animal after it has been slaughtered and the offal, head and hide removed.

Debeak: to cut off a portion of the upper and lower beaks of poultry to prevent cannibalism.

Dehorn: to remove chemically or mechanically the horns of livestock.

Disbud: to remove or prevent growth of the horn buds in young livestock.

Disinfectant: a substance used to kill harmful organisms on non-living surfaces.

Dock: to remove all or part of the tail.

Dodding: removal of soiled wool from a sheep's hind-quarters. (See Dagging)

Doe: a female goat, rabbit or antelope.

Drake: mature male duck.

Drench: to give liquid medicines orally to animals.

Dress: to prepare a carcass for consumption or sale.

Dressing: the covering placed over an open wound to prevent contamination.

Dub: to remove the comb in chicks.

Elastrator: an instrument used to place strong rubber bands over the scrotum or tail for castration or dock-ing, respectively.

Emasculator: an instrument used to sever the cord connec-ting the testes to the body.

Equine: a generic name for horses.

Ewe: a mature female sheep.

Face: to remove the wool from the face region.

Farrow: to give birth to pigs.

Filly: a young female horse.

Flank (n): the part of the animal just in front of the hind leg.

Flank (v): a technique for casting a calf.

Flay: to remove the hide or skin during slaughter.

Fleece: the total wool coat of a sheep.

Flock: a group of animals (sheep, goats, birds). (See Herd)

Flush: to increase feed level of females prior to breeding to increase ovulation rate.

Foal: a young horse of either sex. (See Colt, Filly)

Foggage: grass grown for winter grazing.

Full-mouth: a sheep or bovine that has all its permanent incisor teeth intact and fully developed.

Gander: mature male goose.

Gelding: a castrated male horse.

Gestation: the time period between conception and parturition.

Gilt: a female pig prior to her first litter. (The first litter is sometimes referred to as a 'gilt's litter').

Gimmer: a female sheep prior to her first lambing.

Goatling: a female goat between one and two years of age.

Gobbler: mature male turkey.

Gummer: a sheep with all the incisor teeth missing.

Hair sheep: sheep raised for meat production which have predominantly hair rather than wool.

Handbreeding: to place the male and female together only when the female is on heat. After service occurs, they are again separated.

Hardware: the collection of metal and other foreign material in the rumen.

Hardware disease: the sickness caused by the puncture of the rumen (and often other organs such as the diaphragm and heart) by the hardware in the rumen. Technically known as traumatic reticuloperitonitis.

Hay: dried grass or legumes stored for winter fodder.

Heat: period when the female will accept service by the male; oestrus.

Heifer: female bovine before its first calf.

Hen: a mature avian female.

Herd: a group of animals (cattle, swine, horses). (See Flock)

Hide: the skin of cattle, either raw or dressed.

Hog: a pig (male or female).

Hogg: a castrated male sheep, usually one year old kept for mutton production.

Hogget: see Hogg.

Hook: *tuber coxae*, the hip bones.

Hoven: see Bloat.

Incubate: to hatch eggs, either naturally or artificially by keeping them warm.

Inseminate: to place semen in the female reproductive tract.

Intramuscular: describes injections which are placed in the muscles.

Intravenous: describes injections which are placed into the blood stream by way of a vein.

Joining: placing males with females at the beginning of the breeding season.

Keeling: marking the brisket and chest of a ram with a coloured grease. This colour rubs off on the rump of the ewe at service. (See Raddle)

Kid: a young goat of either sex.

Killing-out percentage: the percentage of the live-weight of an animal which is left after removal of the blood, intestines, hide, head and feet.

Kindle: to give birth to rabbits.

Lactation: the period of time that an animal is in milk.

Lactation interval: the period of time from the beginning of one lactation to the beginning of the next.

Lamb: a young sheep which is suckling; or meat from a young sheep.

Legume: a member of the plant family *Leguminacae*. These have a high protein content and are associated with nitrogen fixation in the soil.

Lick: nutrients available to the animal by licking rather than biting and chewing. Usually includes mineral supplements.

Litter: material placed on the floor to absorb moisture;

or a group of young born to one mother at one time (swine, rabbits).

Livestock: a collective term to denote those animals kept on a farm for productive purposes.

Liveweight: the weight of an animal before it is slaughtered.

Mare: a female horse which has had at least one foal.

Mastitis: a bacterial infection of the udder.

Measle: bladderworm stage of tapeworms found in the active muscles of infected animals.

Moulting: shedding of an outer covering and development of a new one. In poultry is associated with a period of non-productivity.

Mutton: meat from mature sheep.

Natural service: insemination of the female by the male.

Needle teeth: small, sharp teeth, that piglets have at birth - also called 'wolf teeth'.

Nurse: to receive nutrients by sucking the teats of the mother (suckling).

Offal: intestines and other internal parts of a carcass.

Open-faced: sheep without wool on their faces.

Ovine: a generic name for sheep.

Ox: a castrated male bovine used primarily for draught purposes.

Parasite: an organism which lives off another organism (the host) and causes harm to it.

Parturition: the process of giving birth to young.

Pasteurise: to treat by heat to kill microorganisms (usually applies to milk).

Pasting: blockage of the vent of chicks by faeces.

Pig: see Swine.

Placenta: see Afterbirth.

Point-of-lay: the time at which pullets begin to lay eggs. Usually 18 to 20 weeks of age.

Porcine: a generic name for pigs.

Porker: a pig butchered at about 50 kg liveweight or less.

Pullet: female chicken up to the completion of the first

laying season.

Purebred: the offspring of the mating of a male and female of the same breed.

Raddle: to place paint or other marking materials on the male so any females mounted will be marked. (See Keeling)

Ram: a mature, male sheep.

Restrain: to stop the movements of an animal so it can be examined or treated.

Rotational grazing: the regular, cyclic use of pastures to promote grass growth and combat parasite infestation.

Roughage: foodstuffs high in fibre and low in digestible nutrients.

Ruminant: a member of the class of animals having a four-chambered stomach (cattle, sheep, goats, deer, etc.).

Savage: to cause deliberate injury of an animal by another.

Semen: the discharge ejaculated from the testes and accessory sex glands of the male which includes sperm and accessory fluids.

Scour: diarrhoea - may be caused by diet or disease.

Scurf: dried outer skin which will flake off.

Separator: a device used to separate cream from milk.

Service: the insemination of the female.

Shear: to remove the fleece from a sheep.

Shearling: a wool sheep that has had one wool clip removed. It later becomes a two-shear, three-shear, etc. (i.e. one, two and three year old).

Shoat: a young pig.

Sire: the father of an animal.

Silage: green plant materials which have been preserved by fermentation.

Skin (v): to remove the hide or skin of an animal.

Skin (n): the outer covering of animals such as sheep, goats and rabbits.

Small stock: small livestock kept for productive purpose which do not require large facilities or lands, e.g. rabbits, poultry, etc.

Sow: a female pig which has had at least one litter.

Stag: a male castrated after reaching maturity.

Staring: the dry, rough, upstanding coat of an animal which is poorly nourished or diseased.

Stallion: a mature entire male horse.

Steam-up: to give extra feed to a milk-producing animal in the 6-8 weeks prior to parturition.

Steer: a male bovine castrated while young.

Sterilise: to treat instruments etc. with heat or chemicals to destroy harmful organisms.

Stocker: a beef calf between weaning and installation in a feedlot.

Strip cup: a small cup used for checking for mastitis in dairy cows prior to milking.

Stud: a male kept for breeding purposes or a farm where a number of such males are produced and/or kept.

Subcutaneous: describes injections given just under the skin.

Supplement: feed provided to livestock in addition to grazing.

Swine: a collective name for beasts of the genus *Sus* kept for meat purposes.

Syringe: an instrument used to inject materials into an animal either under the skin (subcutaneous), into the muscle (intramuscular), or into the blood stream (intravenous) or, less commonly, by other routes.

Tagging: removal of soiled wool (see Dagging).

Tattoo: an indelible mark made on an animal by forcing ink into wounds make by small needles.

Teat: the appendages, which the young animal suckles, which are attached to, and part of, the mammary gland.

Thermometer: a device used for measuring the body temperature of animals.

Thurl: lateral protrusion of the trochanter major.

Tolly: calf.

Topping: cutting high grass in pastures which helps prevent seeding.

Tup: a male sheep.

Tupping: the act of a ram (tup) serving a ewe.

Tusk: elongated or enlarged permanent canine tooth. Usually seen on boars.

Type (adj): a total of the characteristics which define an animal as being of a specific breed or for a specific purpose.

Type (n): the ideal of body construction that makes an individual best suited for a particular purpose.

Udder: the milk-producing, glandular tissue of the female.

Udder kinch: a rope tied around the body just in front of the hips and udder to prevent kicking.

Vaccinate: the injection of material into an animal to promote long-lasting immunity in the animal or, at least, the ability to tolerate a disease.

Veal: meat from young calves specially fed on an all milk diet.

Veld: grassland areas of southern Africa. Similar to prairies of the USA, pampas (or savannahs) of South America and steppes of Russia.

Vice: a habit or action of animals that is detrimental to themselves or others e.g. cannibalism in poultry or tail-biting in swine.

Wean: to separate the young from the mother so they can no longer suckle.

Wether: a sheep castrated at an early age.

Wool sheep: sheep kept primarily for the production of wool.

Worm: to give medicines to kill worms parasitising the animal.

Yearling: a bovine in its second year of life which has not yet produced young.

Appendix 2: Conversion Factors

LENGTH

1 inch = 25.4 mm.   1 metre = 39.4 in.
1 foot = 0.305 m.
1 yard = 0.914 m.

AREA

$1 \text{ in}^2 = 645 \text{ mm}^2.$ $1 \text{ mm}^2 = 0.0015 \text{ in}^2.$
$1 \text{ ft}^2 = 0.0929 \text{ m}^2.$ $1 \text{ m}^2 = 10.76 \text{ ft}^2.$

VOLUME

1 gallon = 4.546 litres.
$1 \text{ ft}^3 = 0.028 \text{ m}^3.$
44 gallons = 200 litres.

MASS

1 lb = 453.6 gm.
1 ton (long) = 2240 lb = 1016.3 kg.
1 ton (short) = 2000 lb = 909.1 kg.
1 tonne = 2204.6 lb = 1000 kg.
1 kg. = 2.2 lb.

PRESSURE

1 bar is equivalent to 750.1 mm or 29.5 inches of mercury.
1 atmosphere = 1.013 bar.

TEMPERATURE

°C = 5/9 (°F - 32).
°F = 9/5 (°C) + 32.

Appendix 3: Book List for Further Reading

1. LIVESTOCK FACILITIES

Ensminger, M.E. (1977) Animal Science. 7th edition.
   Interstate Printers and Publishers, Inc., Danville,
   Illinois.
Ensminger, M.E. (1978) The Stockman's Handbook. 5th
   edition. Interstate Printers and Publishers, Inc.,
   Danville, Illinois.
Kilgour, R. and Dalton, C. (1983) Livestock Behaviour.
   Granada Publishing, London.
Sainsbury, D. (1983) Animal Health. Granada Publishing,
   London.

2. BEEF CATTLE

Allen, D. and Kilkenny, B. (1980) Planned Beef Production.
   Granada Publishing, London.
Cooper, M. McG. and Willis, M.B. (1979) Profitable Beef
   Production. 3rd edition. Farming Press, Ipswich.
Diggins, R.V. and Bundy, C.E. (1971) Beef Production.
   3rd edition. Prentice-Hall, Englewood Cliffs, N.J.
Ensminger, M.E. (1976) Beef Cattle Science. 5th edition.
   Interstate Printers and Publishers, Inc., Danville,
   Illinois.
Friend, J. and Bishop, D. (1978) Cattle of the World.
   Blandford, Essex.
Goodwin, D.H. (1977) Beef Management and Production.
   Hutchinson, London.

3. DAIRY CATTLE

Barrett, M.A. and Larkin, P.J. (1974) Milk and Beef
   Production in the Tropics. Butterworths, London.
Castle, M.E. and Watkins, P. (1979) Modern Milk Produc-
   tion. Faber, London.
Diggins, R.V. and Bundy, C.E. (1961) Dairy Production.
   2nd edition. Prentice-Hall, Englewood Cliffs, N.J.
Ensminger, M.E. (1980) Dairy Cattle Science. 2nd edition.
   Interstate Printers and Publishers, Inc., Danville,
   Illinois.
Russell, K. and Slater, K. (1980) Principles of Dairy
   Farming. 8th edition. Farming Press, Ipswich.
Russell, K. (revised by S. Williams). (1978) The
   Herdsman's Book. 6th edition. Farming Press,
   Ipswich.

4. CALVES

Roy, J.H.B. (1980) The Calf. 4th edition. Butterworths, London.
Webster, A.J.F. (1983) Calf Husbandry, Health and Welfare. Granada Publishing, London.

5. SHEEP

Cooper, M. McG. and Thomas, R.T. (1979) Profitable Sheep Farming. 4th edition. Farming Press, Ipswich.
Diggins, R.V. and Bundy, C.E. (1958) Sheep Production. Prentice-Hall, Englewood Cliffs, N.J.
Ensminger, M.E. (1970) Sheep and Wool Science. 4th edition. Interstate Printers and Publishers, Inc., Danville, Illinois.
Goodwin, D.H. (1979) Sheep Management and Production. 2nd edition. Hutchinson, London.
Johnston, R.G. (1983) Introduction to Sheep Farming. Granada Publishing, London.
Speedy, A. (1980) Sheep Production. Longmans, London.

6. PIGS

Bundy, C.E. and Diggins, R.V. (1975) Swine Production. 4th edition. Prentice-Hall, Englewood Cliffs, N.J.
Ensminger, M.E. (1970) Swine Science. 4th edition. Interstate Printers and Publishers, Inc., Danville, Illinois.
Goodwin, D.H. (1972) Pig Management and Production. Hutchinson, London.
Pond, W.G. and Maner, J.H. (1974) Swine Production in Temperate and Tropical Environments. W.H. Freeman, San Francisco.
Thornton, K. (1981) Practical Pig Production. 2nd edition. Farming Press, Ipswich.
Whittemore, C. (1980) Pig Production. Longmans, London.
Whittemore, C. and Elsley, F. (1976) Practical Pig Production. Farming Press, Ipswich.

7. POULTRY

Bundy, R.V. and Diggins, C.E. (1960) Poultry Production. Prentice-Hall, Englewood Cliffs, N.J.
Ensminger, M.E. (1980) Poultry Science. 2nd edition. Interstate Printers and Publishers, Inc., Danville, Illinois.
North, Mack, O. (1978) Commercial Chicken Production Manual. 2nd edition. AVI Publishing Co. Inc., Westport, Conn.
Sainsbury, D. (1980) Poultry Health and Management. Granada Publishing, London.

Index